Lecture Notes in Computer Science 4973

Commenced Publication in 1973
Founding and Former Series Editors:
Gerhard Goos, Juris Hartmanis, and Jan van Leeuwen

Elena Marchiori   Jason H. Moore (Eds.)

Evolutionary Computation, Machine Learning and Data Mining in Bioinformatics

6th European Conference, EvoBIO 2008
Naples, Italy, March 26-28, 2008
Proceedings

 Springer

Volume Editors

Elena Marchiori
Radboud University Nijmegen
Institute for Computing and Information Sciences (ICIS)
Toernooiveld 1, 6525 ED Nijmegen, The Netherlands
E-mail: elenam@cs.ru.nl

Jason H. Moore
Dartmouth-Hitchcock Medical Center
HB7937, One Medical Center Dr., Lebanon, NH 03756, USA
E-mail: jason.h.moore@dartmouth.edu

Cover illustration: "Ammonite II" by Dennis H. Miller (2004-2005)
www.dennismiller.neu.edu

Library of Congress Control Number: 2008922956

CR Subject Classification (1998): D.1, F.1-2, J.3, I.5, I.2

LNCS Sublibrary: SL 1 – Theoretical Computer Science and General Issues

ISSN   0302-9743
ISBN-10   3-540-78756-9 Springer Berlin Heidelberg New York
ISBN-13   978-3-540-78756-3 Springer Berlin Heidelberg New York

Springer is a part of Springer Science+Business Media

springer.com

© Springer-Verlag Berlin Heidelberg 2008

Typesetting: Camera-ready by author, data conversion by Scientific Publishing Services, Chennai, India
Printed on acid-free paper   SPIN: 12244836   06/3180   5 4 3 2 1 0

Preface

The field of bioinformatics has two main objectives: the creation and maintenance of biological databases, and the discovery of knowledge from life sciences
data in order to unravel the mysteries of biological function, leading to new
drugs and therapies for human disease. Life sciences data come in the form of
biological sequences, structures, pathways, or literature. One major aspect of
discovering biological knowledge is to search, predict, or model specific information in a given dataset in order to generate new interesting knowledge. Computer
science methods such as evolutionary computation, machine learning, and data
mining all have a great deal to offer the field of bioinformatics. The goal of
the 6th European Conference on Evolutionary Computation, Machine Learning,
and Data Mining in Bioinformatics (EvoBIO 2008) was to bring together experts
from these fields in order to discuss new and novel methods for tackling complex
biological problems.

The 6th EvoBIO conference was held in Naples, Italy on March 26-28, 2008
at the "Centro Congressi di Ateneo Federico II". EvoBIO 2008 was held jointly
with the 11th European Conference on Genetic Programming (EuroGP 2008),
the 8th European Conference on Evolutionary Computation in Combinatorial
Optimisation (EvoCOP 2008), and the Evo Workshops. Collectively, the confer-
ences and workshops were organized under the name Evo* (www.evostar.org).

EvoBIO, held annually as a workshop since 2003, became a conference in
2007, and it is now the premiere European event for those interested in the interface between evolutionary computation, machine learning, data mining, bioinformatics, and computational biology. All papers in this book were presented at
EvoBIO 2008 in response to a call for papers that included topics of interest such
as biomarker discovery, cell simulation and modeling, ecological modeling, flux-
omics, gene networks, biotechnology, metabolomics, microarray analysis, phylogenetics, protein interactions, proteomics, sequence analysis and alignment, and
systems biology. A total of 63 papers were submitted to the conference for peer-
review. Of those, 18 (28.6%) were accepted for publication in these proceedings.

We would first and foremost like to thank all authors who spent time and effort to produce interesting contributions to this book. We would like to thank the
members of the Program Committee for their expert evaluation of the submitted
papers, Jennifer Willies, for her tremendous administrative help and coordina-
tion, and Ivanoe De Falco, Ernesto Tarantino, and Antonio Della Cioppa for
their exceptional work as local organizers. Moreover, we would like to thank the
following persons and institutes: the Naples City Council for supporting the local organization and their patronage of the event, Prof. Guido Trombetti, rector
of the University of Naples "Federico II" and Prof. Giuseppe Trautteur of the
Department of Physical Sciences, for their great support of the local organization, the Instituto Tecnologico de Informatica, Valencia, Spain, for hosting the

Evo* website, Anna Isabel Esparcia-Alcázar for serving as Evo* publicity chair, and Marc Schoenauer and the MyReview team (http://myreview.lri.fr/) for providing the conference review management system and efficient assistance.

Finally, we hope that you will consider contributing to EvoBIO 2009.

February 2008   Elena Marchiori
                                                            Jason H. Moore

Organization

EvoBIO 2008 was organized by Evo* (www.evostar.org).

Program Chairs

Elena Marchiori (Radboud University, Nijmegen, NL)
Jason H. Moore (Dartmouth Medical School in Lebanon, NH, USA)

Steering Committee

David W. Corne	Heriot-Watt University, Edinburgh, UK
Elena Marchiori	Radboud University, Nijmegen, NL
Carlos Cotta	University of Malaga, Spain
Jason H. Moore	Dartmouth Medical School in Lebanon, NH, USA
Jagath C. Rajapakse	Nanyang Technological University, Singapore

Program Committee

Jesus S. Aguilar-Ruiz (Spain)	Katharina Huber (UK)
Francisco J. Azuaje (UK)	Antoine van Kampen (NL)
Wolfgang Banzhaf (Canada)	Jens Kleinjung (UK)
Jacek Blazewicz (Poland)	Mehmet Koyuturk (USA)
Clare Bates Congdon (USA)	Natalio Krasnogor (UK)
Dave Corne (UK)	Bill Langdon (UK)
Carlos Cotta (Spain)	Pietro Lio' (UK)
Federico Divina (Spain)	Michael Lones (UK)
Maggie Eppstein (USA)	Bob MacCallum (UK)
Alex Freitas (UK)	Daniel Marbach (Switzerland)
Gary Fogel (USA)	Elena Marchiori (NL)
Gianluigi Folino (Italy)	Andrew Martin (UK)
Paolo Frasconi (Italy)	Jason Moore (USA)
Franca Fraternali (UK)	Pablo Moscato (Australia)
Rosalba Giugno (Italy)	Vincent Moulton (UK)
Raul Giraldez (Spain)	See-Kiong Ng (Singapore)
Jennifer Hallinan (UK)	Carlotta Orsenigo (Italy)
Lutz Hamel (USA)	Jagath Rajapakse (Singapore)
Jin-Kao Hao (France)	Menaka Rajapakse (Singapore)
Tom Heskes (NL)	Michael Raymer (USA)

Table of Contents

A Hybrid Random Subspace Classifier Fusion Approach for Protein Mass Spectra Classification

Amin Assareh[1], Mohammad Hassan Moradi[1], and L. Gwenn Volkert[2]

[1] Department of Biomedical Engineering, Amirkabir University of Technology, Tehran, Iran
asserah83@googlemail.com, mhmoradi@aut.ac.ir
[2] Department of Computer Science, Kent State University, USA
volkert@cs.kent.edu

Abstract. Classifier fusion strategies have shown great potential to enhance the performance of pattern recognition systems. There is an agreement among researchers in classifier combination that the major factor for producing better accuracy is the diversity in the classifier team. Re-sampling based approaches like bagging, boosting and random subspace generate multiple models by training a single learning algorithm on multiple random replicates or sub-samples, in either feature space or the sample domain. In the present study we proposed a hybrid random subspace fusion scheme that simultaneously utilizes both the feature space and the sample domain to improve the diversity of the classifier ensemble. Experimental results using two protein mass spectra datasets of ovarian cancer demonstrate the usefulness of this approach for six learning algorithms (LDA, 1-NN, Decision Tree, Logistic Regression, Linear SVMs and MLP). The results also show that the proposed strategy outperforms three conventional re-sampling based ensemble algorithms on these datasets.

1 Introduction

Rapid advances in mass spectrometry have led to its use as a prime tool for diagnosis and biomarker discovery [1]. The high-dimensionality-small-sample (HDSS) problem of cancer proteomic datasets is the main issue that plagues and propels current research on protein mass spectra classification [2].

The complexity and subtlety of mass spectra patterns between cancer and normal samples may increase the chances of misclassification when a single classifier is used because a single classifier tends to cover patterns originating from only part of the sample space. Therefore, it would be beneficial if multiple classifiers could be trained in such a way that each of the classifiers covers a different part of the sample space and their classification results were integrated to produce the final classification. Resampling based algorithms such as bagging, boosting, or random forests improve the classification performance by associating multiple base classifiers to work as a "committee" or "ensemble" for decision-making. Any supervised learning algorithm can be used as a base classifier. Ensemble algorithms have been shown to not only increase classification accuracy, but also reduce the chances of overtraining since the committee avoids a biased decision by integrating the different predictions from the individual base classifiers [3]. In recent years a variety of approaches to classifier combination have been applied in the domain of protein mass spectra classification [3-8].

E. Marchiori and J.H. Moore (Eds.): EvoBIO 2008, LNCS 4973, pp. 1–11, 2008.

2 Background

Efforts to improve the performance of classifier combination strategies continue to be an active area of research, especially within the field of bioinformatics as the number of available datasets continues to rapidly increase. It has been empirically shown that the decision made by a set (pool/committee/ensemble/team) of classifiers is generally more accurate than any of the individual classifiers. Both theoretical and empirical research has demonstrated that a good team is one where the individual classifiers in the team are both accurate and make their errors on different parts of the input space. In the other words, one of major factors responsible for improving the performance of a classifier combination strategy is the diversity in the classifier team. There is a consensus among researchers in classifier combination that this diversity issue supersedes the importance of the aggregation method [9]. However, the choice of an appropriate aggregation method can further improve the performance of an ensemble of diverse classifiers.

From the architecture prospective, various schemes for combining multiple classifiers can be grouped into three main categories: 1) parallel, 2) cascading (or serial combination), and 3) hierarchical (tree-like). In the parallel architecture, all the individual classifiers are invoked independently, and their results are then combined by a suitable strategy. Most combination schemes in the literature belong to this category. In the gated parallel variant, the outputs of individual classifiers arc selected or weighted by a gating device before they are combined. In the cascading architecture, individual classifiers are invoked in a linear sequence. The number of possible classes for a given pattern is gradually reduced as more classifiers in the sequence have been invoked. For the sake of efficiency, inaccurate but cheap classifiers (low computational and measurement demands) are considered first, followed by more accurate and expensive classifiers. In the hierarchical architecture, individual classifiers are combined into a structure, which is similar to that of a decision tree classifier. The tree nodes, however, may now be associated with complex classifiers demanding a large number of features. The advantage of this architecture is the high efficiency and flexibility in exploiting the discriminant power of different types of features. Using these three basic architectures, even more complicated classifier combination systems can be constructed [9].

Different combiners expect different types of output from individual classifiers. Lei Xu et al. [10] grouped these expectations into three levels: 1) measurement (or confidence), 2) rank, and 3) abstract. At the confidence level, a classifier outputs a numerical value for each class indicating the belief or probability that the given input pattern belongs to that class. At the rank level, a classifier assigns a rank to each class with the highest rank being the first choice. Rank value cannot be used in isolation because the highest rank does not necessarily mean a high confidence in the classification. At the abstract level, a classifier only outputs a unique class label or several class labels (in which case, the classes are equally good). The confidence level conveys the richest information, while the abstract level contains the least amount of information about the decision being made.

Roughly speaking, building an ensemble based classifier system includes selecting an ensemble of individual classification algorithms, and choosing a decision function for combining the classifier outputs. Therefore, the design of an ensemble classifier system involves two main phases: the design of the classifier ensemble itself and the design of the combination function. Although this formulation of the design problem leads one to think that effective design should address both phases, until recently most design methods described in the literature have only focused on one phase [11].

2.1 Classifier Ensemble Design

So far, two main strategies are discussed in the literature on classifier combination: classifier *selection* and classifier *fusion*. The presumption in classifier selection is that each classifier has expertise in some local area of the feature space. When a feature vector x is submitted for classification, the classifier responsible for the vicinity of x is given the highest authority to label x. Classifier fusion, on the other hand, assumes that all classifiers are equally "experienced" in the whole feature space and the decisions of all of them are taken into account for any x.

Classifier fusion approaches are further divided into resampling-based methods and heterogenous methods. The resampling methods generate multiple models by training a single learning algorithm on multiple random replicates or sub-samples of a given dataset whereas the heterogeneous ensemble methods (also called multistrategy methods) train several different learning algorithms on the same dataset. The approach we describe in this paper is clearly a resampling-based method but differs from the standard resampling-based methods of bagging, boosting, and random forest. In general, resampling-based methods take two perspectives: training a learning algorithm utilizing the same subset of features but different subsets of training data (i.e. Bagging [12] or Boosting [13, 14] or alternatively utilizing the same subset of training data but different subsets of the feature set (i.e. Random Forest or Random Subspace algorithms [15, 16]. Our hybrid approach combines these two perspectives by randomly selecting different subsets of training data *and* randomly selecting different features from a feature set.

2.2 Decision Function Design

In this work we investigate four decision functions to allow evaluation of the impact of different functions on our hybrid approach. The decision functions we investigate are the Majority function, the Weighted Majority function, the Mean function and the Decision Template approach. The 2001 paper by Kuncheva et al. [17] provides an excellent reference on the use of Decision Templates for multiple classifier fusion, including a detailed description of the construction of a soft decision profile for use in ensemble systems.

3 Methods

We have applied our approach to two serum protein mass spectra datasets of ovarian cancer, publicly available from the clinical proteomics program of the national cancer institute website (http://home.ccr.cancer.gov/ncifdaproteomics/ppatterns.asp). The first dataset is "Ovarian 8-7-02" which was produced using the WCX2 protein chip. An upgraded PBSII SELDI-TOF mass spectrometer was employed to generate the spectra, which includes 91 controls and 162 ovarian cancer samples. The second dataset is "Ovarian 4-3-02" prepared by the same chip, but the samples were processed by hand and the baseline was subtracted resulting in the negative intensities seen for some values. The spectra contain 100 control, 100 ovarian cancer and 16 benign samples. Each spectrum of these two datasets includes peak amplitude measurements at 15,154 points defined by corresponding *m/z* values in the range 0–20,000 Da. Figure 1 illustrates the mean spectrums of each dataset.

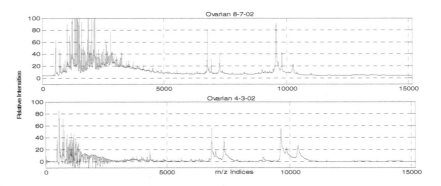

Fig. 1. The mean spectra of the applied datasets: Ovarian 8-7-02 (upper panel) and Ovarian 4-3-02 (lower panel)

Generally, a mass spectrum consists of signals, baseline, and noise. The signals are produced by the peptides, proteins, and contaminants present in the sample; the baseline is the slowly varying trend under the spectrum; and the noise consists of chemical background, electronic noise, signal intensity fluctuations, statistical noise, warping of the signal shapes (due to overcharging in ion traps), and statistical noise in the isotopic clusters (see below). Signals, baseline, and noise can never be totally separated; the baseline, for example, can depend on the presence of large and intense signals as well as on abundant low-intensity noise. Noise can be quite intense and is sometimes impossible to distinguish from real signals. [1]. The goal of preprocessing stage is to ''clean up'' the data such that machine learning algorithms will be able to extract key information and correctly classify new samples based on a limited set of examples [2]. In analyzing mass spectra of blood samples, the preprocessing stage roughly includes three main tasks: baseline correction, smoothing and normalization.

Mass spectra exhibit a monotonically decreasing baseline, which can be regarded as low frequency noise because the baseline lies over a fairly long mass-to-charge ratio range. In this study, we utilized local average within a moving window as a local estimator of the baseline and the overall baseline is estimated by sliding the window over the mass spectrum. The size of the applied window was 200 M/Z. In addition shape preserving piecewise cubic interpolation has been applied to regress the window estimated points to a soft curve. Mass spectra of blood samples also exhibit an additive high frequency noise component. The presence of this noise influences both data mining algorithms and human observers in finding meaningful patterns in mass spectra. The heuristic high frequency noise reduction approaches employed most commonly in studies to date are smoothing filters, the wavelet transform (WT), or the deconvolution filter [2]. We employed a locally weighted linear regression method with a span of 10 M/Z to smooth the spectra. Figure 2 illustrates the smoothing effect on a section of a typical spectrum.

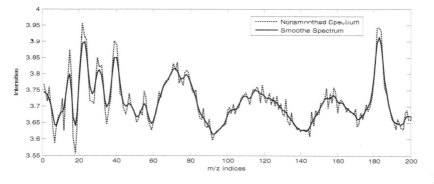

Fig. 2. The effect of the smoothing process on a part of a typical spectrum

A point in a mass spectrum indicates the relative abundance of a protein, peptide or fragment; therefore, the magnitudes of mass spectra cannot be directly compared with each other. Normalization methods scale the intensities of mass spectra to make mass spectra comparable. We normalized the group of mass spectra using total ion current (TIC) method. Figure 3 demonstrates the effect of the preprocessing stages we have applied on a typical mass spectrum from the "Ovarian 8-7-02" dataset.

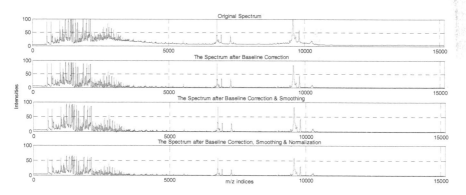

Fig. 3. The effect of preprocessing stages on a typical mass spectrum: The original spectrum (first panel), the spectrum after baseline correction (second panel), the spectrum after baseline correction and smoothing (third panel) and the spectrum after baseline correction, smoothing and normalization (last panel)

3.1 Feature Extraction and Selection

In the present study we use all m/z points as initial features and select the final features set using a t-test with correlation elimination approach. The t-test algorithm with correlation elimination can be succinctly described by the following two steps:

1) Select the first feature based on t-test score as given in equation (1).

$$t = \frac{\left(\overline{x_1} - \overline{x_2}\right)}{\sigma_p \sqrt{\frac{1}{n_1} + \frac{1}{n_2}}} \tag{1}$$

where

$$\sigma_p^2 = \frac{(n_1 - 1)\sigma_1^2 + (n_2 - 1)\sigma_2^2}{n_1 + n_2 - 2} \tag{2}$$

is the pooled standard variance, and $\overline{x_i}$, for $i = 1$ or 2 is the mean of the putative variable in class i, and n_i, for $i = 1$ or 2 is the size of class i.

2) For each of the rest of the potential features, calculate the correlation and local information, w_1 and w_2 respectively, between the applied variable and all previously selected features.

$$w_1 = 1 - R \tag{3}$$

$$w_2 = 1 - e^{-(d/10)^2}$$

where R is the Pearson correlation given in Equation (4),

$$R(x,y) = \frac{Cov(x,y)}{\sqrt{Var(x) \cdot Var(y)}} \tag{4}$$

and d is the distance between the candidate feature and all previously selected features.

From these two steps the score for each feature, designated as *FS*, is then calculated as the product of the t-test and the correlation scores as illustrated in Equation (5).

$$FS = t \times w_1 \times w_2 \tag{5}$$

3.2 Base Learning Algorithms

We test our approach using six well-known base classification algorithms. The following classification algorithms represent a variety of approaches and therefore allow us to assess the robustness of our approach across a variety of classification algorithms. The following learning algorithms have each been applied to the two mass-spectrum data-sets described above as stand alone classifiers using the top 50 features and as base classifiers in our hybrid random subspace fusion ensemble approach.

- Decision Trees
- Linear Discriminant Analysis (LDA)
- 1-Nearest Neighbor (1-NN)

- Logistic Regression
- Linear Support Vector Machines (with a linear kernel)
- Multi Layer Perceptron (MLP) with two hidden layers and 10 neurons in each layer, all the nonlinear functions are tangent-sigmoid and weights were randomly initialized to values in [-1, 1]. The learning function is gradient descent with momentum and back-propagation training was pursued until a limit of 100 epochs or an error of 0 was attained).

3.3 The Proposed Hybrid Random Subspace Classifier Fusion Ensemble Strategy

The heart of our hybrid approach is to randomly choose a subset of training samples and a subset of top features for each of the classifiers that will participate in the ensemble. This approach is hypothesized to maximize the diversity of the ensemble, which has been shown to be an essential feature of effective ensemble approaches. The following steps summarize the proposed strategy for the two-class cases (the strategy can be extended to more cases, but we leave to another paper):

1. Randomly select m samples from the training set (we set m = 60% of training set size)
2. Randomly select n features from n_{max} top-ranked features (we set n =10 and n_{max} = 50)
3. Train a classification algorithm with above selected samples and features
4. Classify the testing samples with the constructed classifier and calculate the corresponding support degree by assigning the *Certainty Factor* (*CF*) to the winner class and (1-*Certinaty Factor*) to the loser class.
5. Iterate above steps for i=1 to I_{max} (we set and I_{max} =100), saving the *CF* for each iteration.
6. Construct a soft decision profile ($I_{max}\times2$) for each test sample using the saved support degrees
7. Inferring the final class from the decision profile using an appropriate decision function. We report in this paper on our experience with Majority, Weighted Majority, Mean, and Decision Template combiners.

4   Results

We compare the performance of our ensemble to each of the base learning algorithms to establish the need for an ensemble in the first place. We then compare the performance of our hybrid random subspace fusion approach to three other well-known resampling based ensemble approaches. For each of the six base classifiers, we selected the 50 top-ranked feature determined by the t-test with correlation elimination as described above. We compared the performance of these base-classifiers to the performance of our proposed hybrid random subspace method on each of the six base learning algorithms for four different decision functions, Majority (MAJ), Weighted Majority (WMAJ), Mean and Decision Template. As described earlier, in each of 100 iterations we randomly select 10 features from the 50 top-ranked features and also randomly selected 60% of the training set. We inferred a *Certainty Factor* for each classifier by testing it over the entire training set and then applied the classifier to the testing samples. After the 100 iterations, we built a soft decision profile for each test using the inferred certainty factor of each classifier. The final classification decision in then determined using one of the four decision templates. This process was repeated 10 times for each choice of base-classifier and decision template in a full 10-fold-cross-validation framework (i.e. 100 total runs for each configuration).

For comparing the classification performance of these different configurations, we used the average of sensitivity and specificity as the performance measure. Although accuracy is the best known measure of classification performance (the number of correctly classified examples over the total number of examples in a given dataset), when class distribution is imbalanced, accuracy can be misleading because it is dominated by performance on the majority class. In two-class problems, accuracy can be replaced by sensitivity and/or specificity. Sensitivity or 'true positive rate' is the number of correctly predicted positive instances over all positive instances. It is the criterion of choice when false negatives incur high penalty, as in most medical diagnosis. Specificity or 'true negative rate' is the number of correctly predicted negative instances over all negative instances. It is used when false alarms are costly [1]. Presentation of the results using this combined measure of sensitivity and specificity allows us to present the results for a large number of different experiments in a relatively small amount of space. Given that overall performance of our approach using this measure is always above 98% we feel this condensed measure is appropriate for this short paper.

Table 1. Performance results obtained on the Ovarian 8-7-02 dataset, for each of six learning algorithms operating either as individual classifiers (utilizing 10 or 50 top features) or as part of the proposed Hybrid Random Subspace strategy utilizing one of four decision functions

Learning Algorithm	Individual Classifier Performance		Hybrid Random Subspace Fusion Ensemble Performance			
	10 Top Features	50 Top Features	Majority	Weighted Majority	Mean	Decision Template
LDA	99.76 (0.2)	**100**	99.98 (0.1)	99.98 (0.1)	99.98 (0.1)	99.98 (0.1)
1-NN	98.95 (.07)	99.23 (0.7)	**100**	**100**	**100**	**100**
Decision Tree	98.36 (0.9)	98.18 (0.5)	**99.9 (0.1)**	**99.9 (0.1)**	**99.9 (0.1)**	**99.9 (0.1)**
Logistic Regression	99.77 (0.3)	99.92 (0.2)	**99.98 (0.1)**	**99.98 (0.1)**	**99.98 (0.1)**	**99.98 (0.1)**
Linear SVMs	99.48 (0.4)	99.89 (0.2)	**99.98 (0.1)**	**99.98 (0.1)**	98.32 (0.6)	**99.98 (0.1)**
MLP	98.46 (2.1)	99.31 (1.8)	**100**	**100**	**100**	**100**

The results are presented as the Mean and Standard Dev. over all runs for each of the two datasets, Ovarian 8 and Ovarian 4 in Tables 1 and 2 respectively. The results clearly show that our propose hybrid random subspace strategy outperforms the performance of each of the six base classifiers tested. For all approaches the Ovarian 8-7-02 data is generally easier to classify, with all approaches achieving average performance above 93%. The second data set, Ovarian 4-3-02, is clearly a more difficult dataset for all of these approaches, yet our hybrid random subspace strategy still achieves higher average performance regardless of the combination function utilized. We can note that overall higher performance is achieved when using the decision template combination function.

Table 2. Performance results obtained on the Ovarian 4-03-02 dataset, for each of six learning algorithms operating either as individual classifiers (utilizing 10 or 50 top features) and operating under the proposed Hybrid Random Subspace strategy utilizing one of four decision functions

Learning Algorithm	Individual Classifier Performance		Hybrid Random Subspace Fusion Ensemble Performance			
	10 Top Features	50 Top Features	Majority	Weighted Majority	Mean	Decision Template
LDA	95.86 (1.2)	96.04 (1.8)	98.98 (0.4)	**98.99 (0.5)**	**98.99 (0.5)**	98.97 (0.5)
1-NN	90.25 (2.0	92.82 (1.3)	99.46 (0.4)	99.66 (0.3)	99.5 (0.4)	**99.82 (0.2)**
Decision Tree	90.76 (2.3)	90.69 (1.1)	99.64 (0.4)	99.73 (0.4)	99.73 (0.4)	**99.83 (0.3)**
Logistic Regression	96.64 (1.3)	96.53 (1.5)	98.88 (0.6)	98.86 (0.6)	**98.92 (0.6)**	98.69 (0.5)
Linear SVMs	95.89 (1.1)	95.3 (1.3)	**98.39 (0.5)**	98.37 (0.6)	98.32 (0.6)	97.53 (0.2)
MLP	96.06 (1.3)	95.63 (0.8)	99.14 (0.3)	99.36 (0.4)	99.36 (0.4)	**99.45 (0.5)**

Table 3. Performance reportred as the Mean and Standard Dev. of the hybrid random subspace fusion strategy and other resampling strategies, using four different decision functions

Dataset	Fusion Strategy	Performance for Different Decision Functions			
		Majority	Weighted Majority	Mean	Decision Template
Ovarian 8-7-02	Hybrid	**99.97 ±0.1**	**99.97 ±0.1**	**99.97 ±0.1**	**99.97 ±0.1**
	Bagging	99.15 ±0.5	98.33 ±0.1	99.15 ±0.5	99.12 ±0.5
	Boosting	99.27 ±0.7	98.06 ±0.8	99.10 ±0.6	98.89 ±0.9
	Random Forest	99.55 ±0.5	99.90 ±0.2	99.88 ±0.2	99.85 ±0.3
Ovarian 4-3-02	Hybrid	**99.64 ±0.1**	**99.73 ±0.4**	**99.73 ±0.4**	**99.83 ±0.3**
	Bagging	95.28 ±1.2	95.27 ±1.3	95.20 ±1.2	95.21 ±1.2
	Boosting	96.87 ±1.7	96.76 ±1.9	96.32 ±1.8	96.93 ±1.8
	Random Forest	93.53 ±1.0	95.83 ±0.7	95.70 ±0.9	96.10 ±0.7

Given that our approach is a resampling strategy, we have also compared the performance with that of three other resampling strategies, including bagging, boosting and random forest. In Table 3 we provide the performance results for each of these other resampling strategies as obtained for the same four combination functions as used above together with decision trees as the base classifier strategy. The results from the hybrid random subspace strategy as reported for decision trees above are also included to facilitate comparisons between these resampling strategies. We note that for other choices of base classifiers (e.g. LDA, Logistics Regression, etc) the performance of the other resampling strategies is generally worse and is therefore not reported here.

5 Conclusion

In this paper, we have described a new hybrid approach for combining sample subspace and feature subspaces when constructing an ensemble of classifiers. We demonstrate the usefulness of our approach on two public datasets of serum protein mass spectra from ovarian cancer research. Following appropriate preprocessing and dimensionality reduction stages, six well-known classification algorithms were utilized as the base classifiers. The results showed a clear enhancement in the performance of the base classifiers when applying the proposed method. Furthermore the performance enhancement was apparent regardless of the decision function used. Future work will investigate how robust this approach is by applying it to other datasets and testing the use of other base classifiers and combination functions.

References

[1] Hilario, M., Kalousis, A., Prados, J., Binz, P.-A.: Data mining for mass spectra-based cancer diagnosis and biomarker discovery. Drug Discovery Today: BioSilico (Elsevier Ltd) 2, 214–222 (2004)

[2] Hilario, M., Kalousis, A., Pellegrini, C., Muller, M.: Processing and Classification of Mass Spectra, Mass Spectrometry Reviews, vol. 25, pp. 409–449 (2006)

[3] Shin, H., Markey, M.K.: A machine learning perspective on the development of clinical decision support systems utilizing mass spectra of blood samples. Journal of Biomedical Informatics 39, 227–248 (2006)

[4] Assareh, A., Moradi, M.H.: A Novel Ensemble Strategy for Classification of Prostate Cancer Protein Mass Spectra. In: 29th IEEE EMBS Annual International Conference (2007)

[5] Bhanot, G., Alexe, G., Venkataraghavan, B., Levine, A.J.: A robust meta-classification strategy for cancer detection from MS data. Proteomics 6, 592–604 (2006)

[6] Vlahou, A., Schorge, J.O., Gregory, B.W., Coleman, R.L.: Diagnosis of Ovarian Cancer Using Decision Tree Classification of Mass Spectral Data. Journal of Biomedicine and Biotechnology 5, 308–314 (2003)

[7] Wu, B., Abbott, T., Fishman, D., McMurray, W., Mor, G., Stone, K., Ward, D., Williams, K., Zhao, H.: Comparison of statistical methods for classification of ovarian cancer using mass spectrometry data. Bioinformatics 19, 1636–1643 (2003)

[8] Yasui, Y.: A data-analytic strategy for protein biomarker discovery: profiling of high-dimensional proteomic data for cancer detection. Biostatistics 4, 449–463 (2003)

[9] Opitz, D., Maclin, R.: Ensemble Methods: An Empirical Study. Journal of Artificial Intelligence Research 26, 169–198 (1999)

[10] Xu, A., Krzyzak, Suen, C.Y.: Methods of Combining Multiple Classifiers and their Applications to Handwriting Recognition. IEEE Trans. on Systems, Man, and Cybernetics 22 (May/June, 1992)

[11] Bunke, H., Kandel, A.: Hybrid Methods in Pattern Recognition. In: Series in Machine Perception and Artificial Intelligence, vol. 47, Word Scientific, Singapore (2002)
[12] Breiman, L.: Bagging Predictors. Machine Learning 24, 123–140 (1996)
[13] Freund, Y., Schapire, R.: Experiments with a new boosting algorithm. In: Thirteenth International Conference on Machine Learning. Bari, Italy, pp. 148–156 (1996)
[14] Schapire, R.: The Strength of Weak Learnability. Machine Learning 5, 197–227 (1990)
[15] Ho, T.K.: The random subspace method for constructing decision forests. IEEE Trans. Pattern Analysis and Machine Intelligence 21, 832–844 (1998)
[16] Breiman, L.: Random Forests. Machine Learning 45, 5–32 (2001)
[17] Kuncheva, L., Bezdek, J., Duin, R.: Decision Templates for Multiple Classifier Fusion: An Experimental Comparison. Pattern Recognition 34(2), 299–314 (2001)

Using Ant Colony Optimization-Based Selected Features for Predicting Post-synaptic Activity in Proteins

Mohammad Ehsan Basiri, Nasser Ghasem-Aghaee,
and Mehdi Hosseinzadeh Aghdam

Department of Computer Engineering, Faculty of Engineering,
University of Isfahan, Hezar Jerib Ave., 81744 Isfahan, Iran
{Basiri,hosseinzadeh}@comp.ui.ac.ir
aghaee@eng.ui.ac.ir

Abstract. Feature Extraction (FE) and Feature Selection (FS) are the most important steps in classification systems. One approach in the feature selection area is employing population-based optimization algorithms such as Particle Swarm Optimization (PSO)-based method and Ant Colony Optimization (ACO)-based method. This paper presents a novel feature selection method that is based on Ant Colony Optimization (ACO). This approach is easily implemented and because of use of a simple classifier in that, its computational complexity is very low. The performance of proposed algorithm is compared to the performance of standard binary PSO algorithm on the task of feature selection in Post-synaptic dataset. Simulation results on Postsynaptic dataset show the superiority of the proposed algorithm.

Keywords: Feature Selection, Ant Colony Optimization (ACO), Particle Swarm Optimization (PSO), Bioinformatics.

1   Introduction

Several parameters can affect the performance of pattern recognition system among which feature extraction and representation of patterns can be considered as some most important ones. Reduction of pattern dimensionality via feature extraction and selection is one of the most essential steps in data processing [1].

Feature Selection (FS) is extensive and it spreads throughout many fields, including document classification, data mining, object recognition, biometrics, remote sensing and computer vision [2]. Given a feature set of size n, the FS problem is to find a minimal feature subset of size m ($m < n$) while retaining a suitably high accuracy in representing the original features.

The objective of feature selection is to simplify a dataset by reducing its dimensionality and identifying relevant underlying features without sacrificing predictive accuracy. By doing that, it also reduces redundancy in the information provided by the selected features. In real world problems FS is a must due to the abundance of noisy, irrelevant or misleading features [3].

E. Marchiori and J.H. Moore (Eds.): EvoBIO 2008, LNCS 4973, pp. 12–23, 2008.

Exhaustive search is the simplest way, which finds the best subset of features by evaluating all the possible subsets. This procedure is quite impractical even for a moderate size feature set. Because the number of feature subset combinations with m features from a collection of n features are $n!/[m!(n-m)!]$ where $m < n$, $m \neq 0$ and the total number of these combinations is $2^n - 2$.

For most practical problems, an optimal solution can only be guaranteed if a monotonic criterion for evaluating features can be found. However, this assumption rarely holds in the real-world [4]. As a result, we must find solutions which would be computationally feasible and represent a trade-off between solution quality and time.

Usually FS algorithms involve heuristic or random search strategies in an attempt to avoid this prohibitive complexity. However, the degree of optimally of the final feature subset is often reduced [3].

Among too many methods which are proposed for FS, population-based optimization algorithms such as Genetic Algorithm (GA)-based method, Particle Swarm Optimization (PSO)-based method and Ant Colony Optimization (ACO)-based method have attracted a lot of attention. These methods attempt to achieve better solutions by application of knowledge from previous iterations.

Particle Swarm Optimization (PSO) comprises a set of search techniques, inspired by the behavior of natural swarms, for solving optimization problems [5]. PSO is a global optimization algorithm for dealing with problems in which a point or surface in an n-dimensional space best represents a solution. Potential solutions are plotted in this space and seeded with an initial velocity. Particles move through the solution space and certain fitness criteria evaluate them. After a while particles accelerate toward those with better fitness values.

Meta-heuristic optimization algorithm based on ant's behavior (ACO) was represented in the early 1990s by M. Dorigo and colleagues [6]. ACO is a branch of newly developed form of artificial intelligence called Swarm Intelligence. Swarm intelligence is a field which studies "the emergent collective intelligence of groups of simple agents" [7]. In groups of insects which live in colonies, such as ants and bees, an individual can only do simple task on its own, while the colony's cooperative work is the main reason determining the intelligent behavior it shows [8].

ACO algorithm is inspired by ant's social behavior. Ants have no sight and are capable of finding the shortest route between a food source and their nest by chemical materials called pheromone that they leave when moving [7].

ACO algorithm was firstly used for solving Traveling Salesman Problem (TSP) [9] and then has been successfully applied to a large number of difficult problems like the Quadratic Assignment Problem (QAP)[10], routing in telecommunication networks, graph coloring problems, scheduling, etc. This method is particularly attractive for feature selection as there seems to be no heuristic that can guide search to the optimal minimal subset every time [3]. On the other hand, if features are represented as a graph, ants will discover best feature combinations as they traverse the graph.

In this paper a new modified ACO-Based feature selection algorithm, ASFS, has been introduced. The classifier performance and the length of selected feature subset are adopted as heuristic information for ACO. Thus, proposed algorithm needs no priori knowledge of features and it is applied to the problem of predicting whether or not a protein has a post-synaptic activity, based on features of protein's primary sequence and finally, the classifier performance and the length of selected feature subset are considered for performance evaluation.

The rest of this paper is organized as follows. Section 2 presents a brief overview of feature selection methods. Ant Colony Optimization (ACO) is described in Sections 3. Section 4 explains the proposed feature selection algorithm. Section 5 reports computational experiments. It also includes a brief discussion of the results obtained and finally the conclusion is offered in the last section.

2   An Overview of Feature Selection (FS) Approaches

Feature selection algorithms can be classified into two categories based on their evaluation procedure [11]. If an algorithm performs FS independent of any learning algorithm (i.e. it is a completely separate preprocessor), then it is included in filter approach (open-loop approach) category. This approach is mostly based on selecting features based on inter-class separability criterion [11]. If the evaluation procedure is tied to the task (e.g. classification) of the learning algorithm, the FS algorithm is a sort of wrapper approach (closed-loop approach). This method searches through the feature subset space using the estimated accuracy from an induction algorithm as a measure of subset suitability. Although wrappers may produce better results, they are expensive to run and can break down with very large numbers of features. This is due to the use of learning algorithms in the evaluation of subsets, some of which can encounter problems while dealing with large datasets [3,12].

The two mentioned approaches are also classified into five main methods which they are Forward Selection, Backward elimination, Forward/Backward Combination, Random Choice and Instance based method. FS methods may start with no feature, all features, a selected feature set or some random feature subset. Those methods that start with an initial subset usually select these features heuristically beforehand. Features are added (Forward Selection) or removed (Backward Elimination) iteratively and in the Forward/Backward Combination method features are either iteratively added or removed or produced randomly thereafter.

The disadvantage of Forward Selection and Backward Elimination methods is that the features that were once selected/eliminated cannot be later discarded/ re-selected. To overcome this problem, Pudil et al. [13] proposed a method to flexibly add and remove features. This method has been called floating search method.

In the wrapper approach the evaluation function calculates the suitability of a feature subset produced by the generation procedure and it also compares that with the previous best candidate, replacing it if found to be better. A stopping criterion is tested in each iteration to determine whether or not the FS process should continue.

Other famous FS approaches are based on the Genetic Algorithm (GA) [14], Simulated Annealing, Particle Swarm Optimization (PSO) and Ant Colony Optimization (ACO) [3,8,15,16,17].

In [15] a hybrid approach has been proposed for speech classification problem. This method has used combination of mutual information and ACO. The hybrid of ACO and mutual information has been used for feature selection in the forecaster [16]. Furthermore, ACO is used for finding rough set reducts in [3] and a new Ant-Miner which used a different pheromone updating strategy has been introduced in [8]. Also, an ACO-Based method has been used in the application of face recognition systems [17] and some surveys of feature selection algorithms are given in [1,18,19].

3   Ant Colony Optimization (ACO)

In the early 1990s, ant colony optimization (ACO) was introduced by M. Dorigo and colleagues as a novel nature-inspired meta-heuristic for the solution of hard combinatorial optimization (CO) problems. ACO belongs to the class of meta-heuristics, which includes approximate algorithms used to obtain good enough solutions to hard CO problems in a reasonable amount of computation time [20].

The Ant System (AS) algorithm is an element of the Ant Colony Optimization (ACO) family of methods [21]. These algorithms are based on a computational paradigm inspired by real ant colonies and the way they function. The underlying idea was to use several constructive computational agents (simulating real ants). A dynamic memory structure, which incorporates information on the effectiveness of previous choices based on the obtained results, guides the construction process of each agent. The behavior of each single agent is therefore inspired by the behavior of real ants.

The paradigm is based on the observation made by ethologists about the medium used by ants to communicate information regarding shortest paths to food by means of pheromone trails. A moving ant lays some pheromone on the ground, thus making a path by a trail of this substance. While an isolated ant moves practically at random (exploration), an ant encountering a previously laid trail can detect it and decide with high probability to follow it and consequently reinforce the trail with its own pheromone (exploitation). What emerges is a form of autocatalytic process through which the more the ants follow a trail, the more attractive that trail becomes to be followed. The process is thus characterized by a positive feedback loop, during which the probability of choosing a path increases with the number of ants that previously chose the same path. The mechanism above is the inspiration for the algorithms of the ACO family [22].

3.1   ACO for Feature Selection

The feature selection task may be reformulated into an ACO-suitable problem. ACO requires a problem to be represented as a graph. Here nodes represent features, with the edges between them denoting the choice of the next feature.

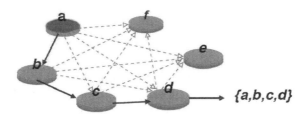

Fig. 1. ACO problem representation for FS

The search for the optimal feature subset is then an ant traversal through the graph where a minimum number of nodes are visited that satisfies the traversal stopping criterion. Figure 1 illustrates this setup. The ant is currently at node a and has a choice of which feature to add next to its path (dotted lines). It chooses feature b next based on the transition rule, then c and then d. Upon arrival at d, the current subset $\{a, b, c, d\}$ is determined to satisfy the traversal stopping criterion (e.g. suitably high classification accuracy has been achieved with this subset). The ant terminates its traversal and outputs this feature subset as a candidate for data reduction. A suitable heuristic desirability of traversing between features could be any subset evaluation function for example, an entropy-based measure [3] or rough set dependency measure [23]. The heuristic desirability of traversal and edge pheromone levels are combined to form the so-called probabilistic transition rule, denoting the probability of ant k at feature i choosing to travel to feature j at time t:

$$P_{ij}^k(t) = \begin{cases} \dfrac{[\tau_{ij}(t)]^\alpha \cdot [\eta_{ij}]^\beta}{\sum\limits_{l \in J_i^k} [\tau_{il}(t)]^\alpha \cdot [\eta_{il}]^\beta} & j \in J_i^k \\ 0 & otherwise \end{cases} \tag{1}$$

Where, η_{ij} is the heuristic desirability of choosing feature j when at feature i (η_{ij} is optional but often needed for achieving a high algorithm performance), J_i^k is the set of neighbor nodes of node i which have not yet been visited by the ant k. $\alpha > 0, \beta > 0$ are two parameters that determine the relative importance of the pheromone value and heuristic information (the choice of α, β is determined experimentally) and $\tau_{ij}(t)$ is the amount of virtual pheromone on edge (i,j).

The overall process of ACO feature selection can be seen in figure 2. The process begins by generating a number of ants, m, which are then placed randomly on the graph i.e. each ant starts with one random feature. Alternatively, the number of ants to place on the graph may be set equal to the number of features within the data; each ant starts path construction at a different feature. From these initial positions, they traverse edges probabilistically until a traversal stopping criterion is satisfied. The resulting subsets are gathered and then evaluated. If an optimal subset has been found or the algorithm has executed a certain number of times, then the process halts and outputs the best feature subset encountered. If none of these conditions hold, then the pheromone is updated, a new set of ants are created and the process iterates once more. The pheromone on each edge is updated according to the following formula:

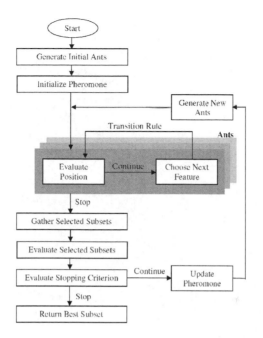

Fig. 2. ACO-based feature selection algorithm

$$\tau_{ij}(t+1) = (1-\rho).\tau_{ij}(t) + \sum_{k=1}^{m} \Delta_{ij}^{k}(t) \qquad (2)$$

where:

$$\Delta_{ij}^{k}(t) = \begin{cases} \gamma'(S^k)/|S^k| & if \ (i,j) \in S^k \\ 0 & otherwise \end{cases} \qquad (3)$$

The value $0 \leq \rho \leq 1$ is decay constant used to simulate the evaporation of the pheromone, S^k is the feature subset found by ant k. The pheromone is updated according to both the measure of the "goodness" of the ant's feature subset ($\acute{\gamma}$) and the size of the subset itself. By this definition, all ants can update the pheromone.

4   Proposed Feature Selection Algorithm

The main steps of proposed algorithm are as follows:

1) Generation of ants and pheromone initialization
   - Determine the population of ants (m).
   - Set the intensity of pheromone trial associated with any feature.
   - Determine the maximum of allowed iterations (T).

2) Ant Foraging and Evaluation
   - Any ant $(k_i, i = 1 : m)$ randomly is assigned to one feature and it should visit all features and build solutions completely. In this step, the evaluation criterion is Mean Square Error (MSE) of the classifier. If an ant is not able to decrease the MSE of the classifier in two successive steps, it will finish its work and exit.
3) Selection of the best ants
   - In this step the importance of the selected subset of each ant is evaluated through classifier performance. Then the subsets according to their MSE are sorted and some of them are selected according to ASFS algorithm.
4) Check the stop criterion
   - Exit, if the number of iterations is more than the maximum allowed iteration, otherwise continue.
5) Pheromone updating
   - For features which are selected in step 3 pheromone intensity are updated.
6) Generation of new ants
   - In this step previous ants are removed and new ants are generated.
7) Go to 2 and continue.

5   Experimental Results

In this section, we report and discuss computational experiments and compare proposed feature selection algorithm with PSO-based approach. The quality of a candidate solution (fitness) is computed by the well-known Nearest Neighbour classifier and the obtained MSE is considered for performance of classifier. Finally, the length of selected feature subset and classifier performance are considered for evaluating the proposed algorithm. For experimental studies we have considered Postsynaptic dataset which is described in the next section.

5.1   Postsynaptic Dataset

This section presents the bioinformatics dataset used in the present work for feature selection. A synapse is a connection between two neurons: presynaptic and postsynaptic. The first is usually the sender of some signals such as the release of chemicals, while the second is the receiver. A post-synaptic receptor is a sensor on the surface of a neuron. It captures messenger molecules from the nervous system, neurotransmitters, and thereby functions in transmitting information from one neuron to another [24]. The dataset used in this paper is called the Postsynaptic dataset. It has been recently created and mined in [24,25]. The dataset contains 4303 records of proteins. These proteins belong to either positive or negative classes. Proteins that belong to the positive class have postsynaptic activity while negative ones don't show such activity. From the 4303 proteins on the dataset, 260 belong to the positive class and 4043 to the negative class. This dataset has many features which makes the feature selection task challenging. More precisely, each protein has 443 PROSITE patterns, or features. PROSITE is a database of protein families and domains. It is based on the observation that, while there are a huge number of different proteins, most of them can be grouped, on the basis of similarities in their sequences, into a limited number of families (a protein consists of a sequence of amino acids). PROSITE patterns are small regions within a protein that present a high sequence similarity when compared to other proteins. In our dataset the absence of a given PROSITE pattern is indicated by a value of 0 for the feature corresponding to that PROSITE pattern which its presence is indicated by a value of 1 for that same feature [25].

5.2   Experimental Methodology

The computational experiments involved a ten-fold cross-validation method [26]. First, the 4303 records in the Postsynaptic dataset were divided into 10 almost equally sized folds. There are three folds containing 431 records each one and seven folds containing 430 records each one. The folds were randomly generated but under the following regulation. The proportion of positive and negative classes in every single fold must be similar to the one found in the original dataset containing all the 4303 records. This is known as stratified cross-validation. Each of the 10 folds is used once as test set and the remaining of the dataset is used as training set. Out of the 9 folds in the training set, one is reserved to be used as a validation set.

In each of the 10 iterations of the cross-validation procedure, the predictive accuracy of the classification is assessed by 3 different methods:

1. Using all the 443 original features: all possible features are used by the Nearest Neighbor classifier.
2. Standard binary PSO algorithm: only the features selected by the best particle found by the binary PSO algorithm are used by the Nearest Neighbor classifier.
3. Proposed ASFS algorithm: only the features selected by the best ant found by the ASFS algorithm are used by the Nearest Neighbor classifier.

As the standard binary PSO and the ASFS algorithms are stochastic algorithms, 20 independent runs for each algorithm were performed for every iteration of the cross-validation procedure. The obtained results, averaged over 20 runs, are reported in Table 1. The average number of features selected by the feature selection algorithms has always been rounded to the nearest integer. The population size (m) used for both algorithms is 30 and the maximum number of iterations (T) equals 50. The standard binary PSO algorithm uses an inertia weight value $w = 0.8$. The choice of the value of this parameter was based on the work presented in [27]. The acceleration constants were $c_1 = c_2 = 2$. For ASFS, various parameters for leading to better convergence are tested and the best parameters that are obtained by simulations are $\alpha = 1$, $\beta = 0.1$ and $\rho = 0.2$. The initial pheromone intensity of each feature is equal to 1.

On our experiments we use a measurement for the accuracy rate of a classification model which has also been used before in [24]. This measurement is given by the equation (4).

$$Predictive\ accuracy\ rate = TPR \times TNR \qquad (4)$$

Where,

$$TPR = \frac{TP}{TP + FN}, \quad TNR = \frac{TN}{TN + FP} \qquad (5)$$

TP, TN, FP and FN are the numbers of true positives, true negatives, false positives and false negatives, respectively [25].

5.3   Comparison of ASFS and Standard Binary PSO

Table 1 gives the optimal selected features for each method. As discussed earlier the experiments involved 200 runs of ACO (ASFS) and standard binary PSO, 10 cross-validation folds times 20 runs with different random seeds. Presumably, those 200 runs selected different subsets of attributes. So, the features which have been listed in table 1 are the ones most often selected by ASFS and standard binary PSO across all the 20 runs. Both ACO-Based and PSO-Based methods significantly reduce the number of original features, however; ACO-Based method, ASFS, chooses fewer features.

Table 1. Selected Features of standard binary PSO and ASFS algorithms

Method	Selected Features	Number of Selected Features
Binary PSO	134, 162, 186, 320, 321 333, 342, 351,352, 353	10
ACO (ASFS)	352, 381, 419, 353, 342	5

Also, the results of both algorithms for all of the 10 folds are summarized in table 2. The classification quality and feature subset length are two criteria which are considered to assess the performance of algorithms. Comparing these criteria, we noted that ASFS and standard binary PSO algorithms did very better than the baseline algorithm (using all features). Furthermore, for all of the 10 folds the ASFS algorithm selected a smaller subset of features than the standard binary PSO algorithm.

As we can see in table 2, the average number of selected features for standard binary PSO algorithm was equal to 15.4 with the average predictive accuracy of 0.77 and the average number of selected features for ASFS algorithm was equal to 5.1 with the average predictive accuracy of 0.84. Furthermore, in [25] a new discrete PSO algorithm, called DPSO, has been introduced for feature selection. DPSO has been applied to Postsynaptic dataset and the average number of features selected by that was 12.70 with the average predictive accuracy of 0.74.

Table 2. Comparison of obtained results for proposed algorithm and standard binary PSO

Fold	Using all the 443 original attribute			Standard Binary PSO algorithm				Proposed ASFS Algorithm			
	TPR	TNR	TPR × TNR	TPR	TNR	TPR × TNR	No. of Features selected	TPR	TNR	TPR × TNR	No. of Features selected
1	1.00	1.00	1.00	1.00	1.00	1.00	16	1.00	1.00	1.00	6
2	1.00	1.00	1.00	1.00	1.00	1.00	19	1.00	1.00	1.00	2
3	1.00	1.00	1.00	1.00	1.00	1.00	21	1.00	1.00	1.00	4
4	0.73	1.00	0.73	0.76	1.00	0.76	14	0.73	1.00	0.73	3
5	0.00	1.00	0.00	0.69	0.92	0.63	17	0.73	0.96	0.70	5
6	0.00	1.00	0.00	0.65	0.96	0.62	18	0.88	0.99	0.87	5
7	0.92	1.00	0.92	0.88	1.00	0.88	11	0.92	1.00	0.92	9
8	1.00	1.00	1.00	0.69	1.00	0.69	15	1.00	1.00	1.00	6
9	0.73	1.00	0.73	0.73	1.00	0.73	9	0.73	1.00	0.73	3
10	0.42	1.00	0.42	0.42	1.00	0.42	14	0.42	1.00	0.42	8
AVG	0.68	1.00	0.68	0.78	0.99	0.77	15.4	0.84	0.99	0.84	5.1

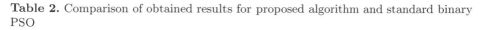

Fig. 3. (a) Accuracy rate of feature subsets obtained using ASFS and standard binary PSO. (b) Number selected features.

Comparison of these three algorithms shows that ASFS tends to select a smaller subset of features than the standard binary PSO algorithm and DPSO. Also, the average predictive accuracy of ASFS is higher than that of the standard binary PSO algorithm and DPSO. Predictive accuracy and number of selected features for ASFS and standard binary PSO algorithm are shown in figure 3.

6 Conclusion

In this paper a novel ACO-Based feature selection algorithm, ASFS, is presented. In the proposed algorithm, the classifier performance and the length of selected feature subset are adopted as heuristic information for ASFS. So, we can select the optimal feature subset without the prior knowledge of features. To show the utility of proposed algorithm and to compare it with a PSO-Based approach a set of experiments was carried out on Postsynaptic dataset. The computational results indicate that ASFS outperforms standard binary PSO method since it achieved better performance with the lower number of features.

Acknowledgments. The authors wish to thank the Office of Graduate studies of the University of Isfahan for their support and also wish to offer their special thanks to Dr. Alex A. Freitas for providing Postsynaptic dataset.

References

1. Kml, L., Kittler, J.: Feature set search algorithms. In: Chen, C.H. (ed.) Pattern Recognition and Signal Processing, Sijhoff and Noordhoff, The Netherlands (1978)
2. Ani, A.A.: An Ant Colony Optimization Based Approach for Feature Selection. In: Proceeding of AIML Conference (2005)
3. Jensen, R.: Combining rough and fuzzy sets for feature selection. Ph.D. Thesis, University of Edinburgh (2005)
4. Kohavi, R.: Feature Subset Selection as search with Probabilistic Estimates. In: AAAI Fall Symposium on Relevance (1994)
5. Kennedy, J., Eberhart, R.C.: Swarm Intelligence. Morgan Kaufmann Publishers Inc., San Francisco (2001)
6. Dorigo, M., Caro, G.D.: Ant Colony Optimization: A New Meta-heuristic. In: Proceeding of the Congress on Evolutionary Computing (1999)
7. Bonabeau, E., Dorigo, M., Theraulaz, G.: Swarm Intelligence: From Natural to Artificial Systems. Oxford University Press, New York (1999)
8. Liu, B., Abbass, H.A., McKay, B.: Classification Rule Discovery with Ant Colony Optimization. IEEE Computational Intelligence 3(1) (2004)
9. Dorigo, M., Maniezzo, V., Colorni, A.: The Ant System: Optimization by a Colony of Cooperating Agents. IEEE Transactions on Systems, Man, and Cybernetics, Part B 26(1), 29–41 (1996)
10. Maniezzo, V., Colorni, A.: The Ant System Applied to the Quadratic Assignment Problem. Knowledge and Data Engineering 11(5), 769–778 (1999)
11. Duda, R.O., Hart, P.E.: Pattern Recognition and Scene Analysis. Wiley, Chichester (1973)
12. Forman, G.: An extensive empirical study of feature selection metrics for text classification. Journal of Machine Learning Research 3, 1289–1305 (2003)
13. Pudil, P., Novovicova, J., Kittler, J.: Floating search methods in feature selection. Pattern Recognition Letters 15, 1119–1125 (1994)
14. Siedlecki, W., Sklansky, J.: A note on genetic algorithms for large-scale feature selection. Pattern Recognition Letters 10(5), 335–347 (1989)
15. Ani, A.A.: Ant Colony Optimization for Feature Subset Selection. Transactions on Engineering, Computing and Technology 4 (2005)
16. Zhang, C.K., Hu, H.: Feature Selection Using the Hybrid of Ant Colony Optimization and Mutual Information for the Forecaster. In: Proceedings of the Fourth International Conference on Machine Learning and Cybernetics (2005)
17. Kanan, H.R., Faez, K., Hosseinzadeh, M.: Face Recognition System Using Ant Colony Optimization-Based Selected Features. In: Proceeding of the First IEEE Symposium on Computational Intelligence in Security and Defense Applications. CISDA 2007, pp. 57–62. IEEE Press, USA (2007)

18. Bins, J.: Feature Selection of Huge Feature Sets in the Context of Computer Vision. Ph.D. Dissertation, Computer Science Department, Colorado State University (2000)
19. Siedlecki, W., Sklansky, J.: On Automatic Feature Selection. International Journal of Pattern Recognition and Artificial Intelligence 2(2), 197–220 (1988)
20. Dorigo, M., Blum, C.: Ant colony optimization theory: A survey. Theoretical Computer Science 344, 243–278 (2005)
21. Dorigo, M., Di Caro, G., Gambardella, L.M.: Ant algorithms for discrete optimization. Artificial Life 5, 137–172 (1999)
22. Engelbrecht, A.P.: Fundamentals of Computational Swarm Intelligence. Wiley, London (2005)
23. Pawlak, Z.: Rough Sets: Theoretical Aspects of Reasoning about Data. Kluwer Academic Publishing, Dordrecht (1991)
24. Pappa, G.L., Baines, A.J., Freitas, A.A.: Predicting post-synaptic activity in proteins with data mining. Bioinformatics 21(2), 19–25 (2005)
25. Correa, E.S., Freitas, A.A., Johnson, C.G.: A new discrete particle swarm algorithm applied to attribute selection in a bioinformatics dataset. In: The Genetic and Evolutionary Computation Conference - GECCO-2006, Seattle, pp. 35–42 (2006)
26. Witten, I.H., Frank, E.: Data Mining: Practical Machine Learning Tools and Techniques. Morgan Kaufmann, San Francisco (2005)
27. Shi, Y., Eberhart, R.C.: Parameter selection in particle swarm optimization. In: Porto, V.W., Waagen, D. (eds.) EP 1998. LNCS, vol. 1447, pp. 591–600. Springer, Heidelberg (1998)

Generating Linkage Disequilibrium Patterns in Data Simulations Using genomeSIMLA

Todd L. Edwards[1], William S. Bush[1], Stephen D. Turner[1], Scott M. Dudek[1],
Eric S. Torstenson[1], Mike Schmidt[2], Eden Martin[2], and Marylyn D. Ritchie[1]

[1] Center for Human Genetics Research, Department of Molecular Physiology & Biophysics,
Vanderbilt University, Nashville, TN, USA
{edwards,bush,stephen,dudek,torstenson,
ritchie}@chgr.mc.vanderbilt.edu
[2] Center for Statistical Genetics and Genetic Epidemiology, Miami Institute for Human
Genetics, University of Miami Miller School of Medicine, Miami, FL, USA
{mschmidt,emartin1}@med.miami.edu
http://chgr.mc.vanderbilt.edu/genomeSIMLA

Abstract. Whole-genome association (WGA) studies are becoming a common
tool for the exploration of the genetic components of common disease. The
analysis of such large scale data presents unique analytical challenges, including problems of multiple testing, correlated independent variables, and large
multivariate model spaces. These issues have prompted the development of
novel computational approaches. Thorough, extensive simulation studies are a
necessity for methods development work to evaluate the power and validity of
novel approaches. Many data simulation packages exist, however, the resulting
data is often overly simplistic and does not compare to the complexity of real
data; especially with respect to linkage disequilibrium (LD). To overcome this
limitation, we have developed genomeSIMLA. GenomeSIMLA is a forward-
time population simulation method that can simulate realistic patterns of LD in
both family-based and case-control datasets. In this manuscript, we demonstrate
how LD patterns of the simulated data change under different population
growth curve parameter initialization settings. These results provide guidelines
to simulate WGA datasets whose properties resemble the HapMap.

Keywords: Whole genome association, data simulation, linkage disequilibrium.

1 Introduction

The initial success of the human genome project is the nearly complete characterization
of the consensus human sequence. This has greatly increased our ability to describe the
structure of genes and the genome and to better design experiments. Perhaps of even
more importance for disease gene studies is the continuously updated HapMap data [1].
This vast pool of characterized common differences between individuals greatly increases our ability to perform targeted or whole genome association (WGA) studies by
using the measured patterns of linkage disequilibrium (LD) as a foundation for single
nucleotide polymorphism (SNP) selection and data interpretation. SNPs are single base

E. Marchiori and J.H. Moore (Eds.): EvoBIO 2008, LNCS 4973, pp. 24–35, 2008.

changes in DNA that vary across individuals in a population at a measurable frequency. WGA studies interrogate hundreds of thousands of SNPs throughout the entire human genome in an effort to map disease susceptibility or drug response to common genetic variation.

LD is the nonrandom association of alleles at multiple SNPs. This association can be quantified by the squared Pearson's product-moment correlation coefficient (R^2). Also available is a related measure, D', which is the proportion of the maximum possible R^2 given a difference in allele frequencies. The R^2 value gives an indication of the statistical power to detect the effect on disease risk of an ungenotyped SNP, whereas D' is indicative of past recombination events.

Advances that increase the complexity of data simulations will permit investigators to better assess new analytical methods. GenomeSIMLA (an extension of [2]) was developed for the simulation of large-scale genomic data in population based case-control or family-based samples. It is a forward-time population simulation algorithm that allows the user to specify many evolutionary parameters to control evolutionary processes. GenomeSIMLA simulates patterns of LD representative of observed human LD patterns through realistic processes of mating and recombination. This tool will enable investigators to evaluate the sampling properties of any statistical method which is applied to large-scale data in human populations. We describe the algorithm and demonstrate its utility for future genetic studies with WGA.

1.1 Background

Multiple technologies now allow a WGA design to be implemented by genotyping between 500,000 and 1.8 million SNPs with high fidelity and low cost. It is conceivable that technological advances will lead to whole genome sequencing in the not too distant future that will involve generating 10-20 million base pair variations per individual. In a WGA approach, a dense map of SNPs is genotyped and alleles, genotypes, or haplotypes are tested directly for association with disease. Estimates suggest that with 500,000 SNPs, ~50-75% of the common variation in the genome is captured. Recent studies have shown that the precise extent of coverage is dependent on study design, population structure, and allele frequency [3]. Regardless, WGA is by far the most detailed and complete method of genome interrogation currently possible. WGA has successfully detected association with genetic variation in several common diseases including breast cancer [4,5], type II diabetes [6-9], obesity [10], and others [6].

1.2 GenomeSIM and SIMLA

GenomeSIM [2] was developed for the simulation of large-scale genomic data in population based case-control samples. It is a forward-time population simulation algorithm that allows the user to specify many evolutionary parameters and control evolutionary processes. SIMLA (or SIMulation of Linkage and Association) [11,12] is a simulation program that allows the user to specify varying levels of both linkage and LD among and between markers and disease loci.

SIMLA was specifically designed for the simultaneous study of linkage and association methods in extended pedigrees, but the penetrance specification algorithm can also be used to simulate samples of unrelated individuals (e.g., cases and controls). We have combined genomeSIM as a front-end to generate a population of founder chromosomes. This population will exhibit the desired patterns of LD that can be used as input for the SIMLA simulation of disease models. Particular SNPs may be chosen to represent disease loci according to desired location, correlation with nearby SNPs, and allele frequency. Using the SIMLA method of disease modeling, up to six loci may be selected for main effects and all possible 2 and 3-way interactions as specified in [13] among these 6 loci are available to the user as elements of a disease model. Once these loci are chosen the user specifies disease prevalence, a mode of inheritance for each locus, and relative risks of exposure to the genotypes at each locus. An advantage of the SIMLA approach to the logistic function is it can simulate data on markers that are not independent, yet yield the correct relative risks and prevalence. Many simulation packages using a logistic function for penetrance specification do not have this capability. Modeling of purely epistatic interactions with no detectable main effects, as in genomeSIM, is also supported separately and can simulate 2-way, 3-way, up to n-way interactions. Purely epistatic modeling allows the user to specify a model odds ratio, heritability, and prevalence for disease effects. Thus, the marriage of genomeSIM and SIMLA has allowed for the simulation of large scale datasets with realistic patterns of LD and diverse realistic disease models in both family-based and case-control data. We describe the new software package, genomeSIMLA, in section 2.2.

1.3 Alternative Genetic Data Simulation Packages

Several genetic data simulation packages are currently available. SIMLINK [14,15], SIMULATE, and SLINK [16] will simulate pedigrees from an existing dataset. Coalescent-based methods [17] have been used for population based simulation in genetic studies; however, standard approaches which are extremely efficient in simulating short sequences, are not successful for long sequences. GENOME is a novel coalescent-based whole genome simulator developed to overcome previous limitations [18]. HAP-SAMPLE uses the existing Phase I/II HapMap data to resample existing phased chromosomes to simulate datasets [19]. In recent years, forward-time population simulations have been developed including easyPOP [20], FPG [21], FREGENE [22], and simuPOP [23]. All of the existing simulation packages have strengths and weaknesses. The motivation for developing genomeSIMLA is to achieve the ability to simulate: 1) realistic patterns of LD in human populations, 2) WGA datasets in both family and case-control study designs, 3) single or multiple independent main effects, and 4) purely epistatic gene-gene interactions in efficient, user friendly software. Existing simulation packages can do one or more of these, but few are able to succeed in all areas.

2   Methods

2.1   GenomeSIMLA

GenomeSIMLA generates datasets using a forward-time population simulator which relies on random mating, genetic drift, recombination, and population growth to allow a population to naturally obtain LD features. An initial population (or pool of chromosomes) is generated using allele frequencies and positions for a set of desired SNPs or random allele frequencies for real or synthetic SNP locations. Recombinant gametes are created based on intermarker recombination probabilities calculated using Kosambi (accounting for recombination interference) or Haldane (accounting for multiple events between polymorphisms) genetic mapping functions. Recombination probability between two polymorphisms is determined by the Kosambi or Haldane function of the map distance based on a 1 centimorgan per 1 million bases genetic map. The number of crossover events for a pair of parental chromosomes to generate gametes is a random Poisson variable where the expected number of events is the sum of all intermarker recombination probabilities for the chromosome. The two resulting gametes, one from each parent, are then combined to create a new individual. The mapping approximation of 1 million bases per centimorgan is applied here; however, other values could be applied to simulate population-specific genetic maps or recombination hotspots. The random mating and recombination process continues on the pool of chromosomes for a set number of generations to generate realistic patterns of LD and produce sufficient numbers of chromosomes for drawing datasets. After the pool of chromosomes has developed suitable LD and grown to a useful size, datasets can be drawn by randomly sampling chromosomes with replacement to create nonredundant individuals. Disease-susceptibility effects of multiple genetic variables can be modeled using either the SIMLA logistic function [11,12] or a purely epistatic multi-locus penetrance function [24] found using a genetic algorithm. These individuals are either mated to yield pedigrees, for family-based datasets, or are evaluated by a logistic function or a purely epistatic penetrance function of their genotypic exposures to determine disease status for case-control datasets.

Figure 1 illustrates the general steps involved in producing a simulated dataset. As a first step, genomeSIMLA establishes the size of the genome based on the user specified parameters. The total number of SNPs is unlimited except by hardware considerations. We are currently able to simulate at least 500K SNPs. The simulator generates the number of SNPs, recombination fraction, and allele frequencies within user specified margins or boundaries. GenomeSIMLA then generates an initial population (or pool of chromosomes) based on the genome established in the previous step. For each SNP in the genome, the simulator randomly assigns an allele to each chromosome based on the allele frequencies of the SNP. A dual-chromosome representation is used for creating individuals to allow for an efficient representation of the genome and for crossover between chromosomes during the mating process. The genotype at any SNP can be determined simply by adding the values of the two chromosomes at that position. As a result, the genotypes range from 0 to 2 at any SNP.

The initial population forms the basis for the second generation in the simulation. For each cross, four chromosomes are randomly selected with replacement to create two individuals to be the parents for a member of the new generation. Each parent contributes one haploid genome to the child. GenomeSIMLA creates the gametic genotype by recombining the parent's chromosomes. The total number of chromosomes in the pool can be constant or follow a population growth model (linear, exponential, or logistic). This will determine the number of mating/crossover events that occur. GenomeSIMLA continues through a specified number of generations depending on the desired LD patterns.

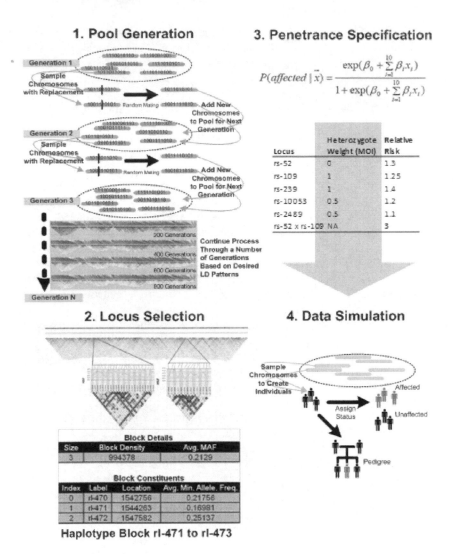

Haplotype Block rl-471 to rl-473

Fig. 1. Simulator Overview. This figure demonstrates the steps of the genomeSIMLA algorithm for simulating data as described in the text. In summary, the process of simulating data is as follows:

1. Develop the chromosome pool using either artificial intermarker distances and recombination or positions from real data. Set the parameters of the population growth to fit the desired LD properties.
2. Select loci to be the disease susceptibility loci in the simulation. Loci can be searched for using built-in search tools allowing the user to screen loci based on allele frequency, block size, and position.
3. Specify the disease model. Either multiple loci with main effects and interactions among them or purely epistatic effects can be modeled
4. Simulate data by either drawing individuals for case-control data or founders for family data.

To create datasets, chromosomes are sampled from the pool with replacement and affection status is assigned based on the user-specified penetrance table or logistic function. Samples are drawn until the desired number of cases and controls are accumulated. In family-based simulation, founders are drawn and offspring are created using the same methods as applied in the pool generation steps. The penetrance function is applied to the parents and offspring to determine status and the resulting pedigrees are retained in a dataset if the study ascertainment criteria are met. Otherwise the pedigrees are discarded and the founder chromosomes are allowed to be drawn again.

2.2 GenomeSIMLA: Implementation

Performance on desktop grade hardware and interpretable results reporting were main goals of software development. Users can simulate data on modern desktop hardware and have their datasets within 24-48 hours for many parameter settings; though the exact time will be dependent upon the particular growth curve used and the desired chromosome pool size. To achieve these goals we focused on memory requirements, threading, and LD plotting.

C++ allows us to utilize memory with minimal overhead; however, retaining 100,000 chromosomes of 500,000 SNPs each is not a trivial task. To maintain this within the limits of a modern desktop machine, we represent each chromosome as a binary string. Also, unless otherwise specified, genomeSIMLA will only have a single chromosome pool in memory. One drawback of using the binary string for a chromosome is that we are limiting genomeSIMLA to biallelic data. By retaining a single pool in memory, our memory requirements fall reasonably under 2 gigabytes of RAM.

We have implemented two different threading mechanisms to allow users to take full advantage of the hardware available to them. When using genomeSIMLA in 32bit environments, there are at most 4 Gigabytes of memory available to the system. To accommodate users with multiple processors running 32bit operating systems, we allow specification of the number of threads per chromosome. This incurs a minimal memory increase but can speed the calculations up considerably. However, when running genomeSIMLA under 64bit, we allow for configurations to specify any number of chromosomes be managed simultaneously. This is limited by available hardware and process time scales almost linearly with the number of processors available.

In order to address our reporting needs, we implemented our own LD plotter. Existing LD plotting software could not accommodate whole chromosome data. As a result, genomeSIMLA is capable of generating whole chromosome LD plots similar to those generated by other software packages. Calculating whole-genome LD statistics on large chromosomal pools is a computationally intensive process. To reduce computation time, LD statistics can be optionally calculated on a sample of the entire pool.

2.3 Growth Curve Parameter Sweep

To develop an understanding of the consequences of different population growth curve parameter settings, we have designed a series of experiments. The hypothesis is that some combination of population growth parameters will emulate the average profile of correlation by distance observed in the HapMap data. We used a generalized logistic curve, or Richards curve, to model realistic population growth [25] Equation 1. The

Richards growth curve consists of five parameters: A -- the initial population size or lower asymptote, C -- the carrying capacity of the population or the upper asymptote, M -- the generation of maximal growth, B – the growth rate, and T – a parameter that determines if the point of maximal growth occurs near the lower or upper asymptote.

$$Y = A + \frac{C}{(1 + Te^{-B(x-M)})^{1/T}} \tag{1}$$

This function provides a parameterized theoretical basis for population growth, though real population growth likely has more stochastic variability. To allow variability in population growth, we implemented a jitter parameter that draws a random number from a uniform distribution over a range specified by the user and adds or subtracts that percentage from the population size predicted by the growth curve. For the purposes of the parameter exploration in this study, however, the jitter parameter was set to zero.

We scanned through a wide range of parameters to find population growth profiles providing suitable correlation among genetic variables for data simulation. Since there were 5 parameters to vary and many possible values for each, we were still limited to a small subset of the possible sets of growth parameters available. Prior to this study, we performed a number of parameter sweeps to evaluate ranges that were likely to yield interesting and realistic LD patterns (results not shown) in a population of 100,000 chromosomes. For this study, we split the parameter sweep into three scans. In total, 726 combinations of parameter settings were examined for average LD over distance.

Table 1. Parameter sweep of population growth parameters for the logistic function: settings for three scans

Parameters	Scan 1	Scan 2	Scan 3
A - Lower asymptote	500, 750, 1000	100, 150, 200, 250, 300	750, 1000, 1250, 1500
C - Upper asymptote	120k, 500k, 900k	110k, 120k	120k
M - Maximum growth time	305, 315, 325, 335, 345, 355	350, 400, 450	500, 1000, 1500, 2000, 2500, 3000
B - Growth rate	0.005, 0.0075, 0.01	0.018, 0.02, 0.022, 0.025	0.02, 0.025, 0.03, 0.035, 0.04
T - Maximum growth position	0.1. 0.2, 0.3	0.1	0.1
Total parameters	486	120	120

We predict that a common usage of genomeSIMLA software will be to simulate case-control and family-based whole-genome association datasets containing 300,000-500,000 biallelic markers across the genome. These data could be used to evaluate the sampling properties of new or established association methods or techniques to characterize the genetic structure of populations. While genomeSIMLA can simulate data of this magnitude, for this study, we wanted to focus on a single chromosome. Thus, we simulated the 6031 chromosome 22 markers used on the popular Affymetrix 500K SNP Chip.

To visualize the results of each parameter combination, average R^2 by distance in kilobases was graphed for the simulated data and for the CEPH (Caucasian), Yoruba (African), and Chinese/Japanese HapMap populations. This representation captures global estimates of correlation by distance across the entire chromosome.

3 Results

Parameter settings in Scan 1 did not yield LD which was comparable to HapMap samples. A trend was observed among the better fitting models that the parameters C and T always functioned best when set to 120k and 0.1, respectively. Scan 2 examined very small initial populations and more rapid growth to strengthen LD profiles through rapid genetic drift. These unfortunately also resulted in fixing many alleles. Scan 3 focused on larger initial populations, late maximum growth, and rapid growth. These simulations were the most successful and resulted in several curves which approximated LD in HapMap samples well. One such example is presented in Figure 2.

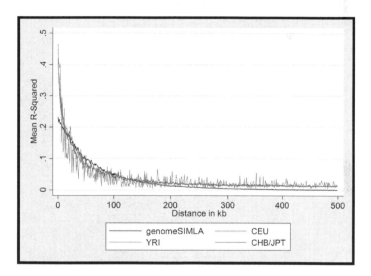

Fig. 2. Average R^2 by distance (kb) for simulated, CEPH (Caucasian), YRI (Yoruba African), and CHB/JPT (Chinese/Japanese) samples

While not a perfect fit to any population, the curve represents a good approximation of the correlation observed in the data. Of note is the fit in the shorter ranges, since short range LD is more related to the power to detect associations with disease susceptibility loci [26]. A sample of the actual LD observed among the markers is presented in Figure 3. The goal of this study was to obtain data which on average is similar to HapMap data. Since we initialized the chromosomes with random minor allele frequency and the measure R^2 is sensitive to this parameter, it is not expected that each intermarker correlation will be identical to the value calculated from the HapMap data. However, it can be seen here that the major features and regions of high and low

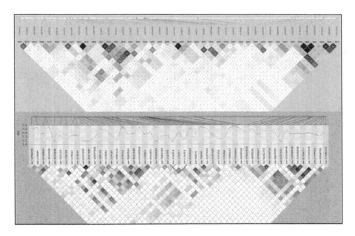

Fig. 3. Sample of LD from the simulation detailed in Figure 2 of R^2 plots from HapMap CEPH samples (above) and simulated data

Table 2. Time to completion for pool advancement to 100,000 chromosomes and graphical LD calculation and representation for up to 500,000 SNPs

Simulation	Processors	LD Calculation	Time
Chr1	1	Sampled	13m 41s
Chr1	1	Complete	88m 45s
Chr1	4	Sampled	5m 41s
Chr1	4	Complete	33m 4s
Chr22	1	Sampled	2m 15s
Chr22	1	Complete	12m 27s
Chr22	4	Sampled	1m 33s
Chr22	4	Complete	4m 30s
500k	4	Sampled	74m 52s
500k	4	Complete	367m 54s
500k	8	Sampled	29m 22s
500k	8	Complete	123m 21s

correlation are captured. The growth curve in Figure 2 and the LD shown in Figure 3 were generated with the following parameters: A=1500, C=120000, M=500, B=0.02, T=0.1. D', an alternate measure of LD, was more difficult to fit than R^2. The curves for the simulated data generally were stronger than those observed for the real data in the short ranges but weaker at long ranges. The reasons for this are unknown but are a topic of further study for genomeSIMLA.

We also measured the time to completion for various size simulations. We examined the markers for the Affymetrix 500K in chromosomes 1 and 22 and the full chip (Table 2) for the growth parameters in Figures 2 and 3. In order to reduce the time required to scan a growth curve for ideal LD patterns, genomeSIMLA utilizes both sampled and complete LD. When generating sampled LD plots, genomeSIMLA draws LD plots for a small region (1000 SNPs) of each chromosome and limits the participants to a relatively small number (1500).

4 Discussion

We found that tuning the parameters to emulate the average pattern of correlation in real human populations was difficult. However, some settings we used provided good qualitative fit to the observed real data. Statistical evaluation of these distributions was difficult, since tests on distributions are extremely sensitive and strong ceiling effects were observed. We initialized our chromosome pools with random allele frequency independent data, and only allowed the recombination probabilities to directly mimic those expected from the Kosambi function for the HapMap physical distances. This procedure was a proof of principle that it is neither necessary to directly resample HapMap chromosomes or use computationally inefficient coalescent models to effectively simulate the properties of unobserved samples from real human populations.

One potential reason for the deviations of our simulated data from those observed in the HapMap populations was that genomeSIMLA simulates phased chromosomes with no heterozygote ambiguity. As a result, genomeSIMLA does not employ an Expectation Maximization (EM) algorithm [27] to phase genotype data. The phased data available from the HapMap is processed from raw genotypes using PHASE[28], which is a notably different means of LD calculation. The effects of EM algorithms on the observed average LD by distance when the true LD is known has not been investigated, but will be a topic of further study.

The results we observed here show that genomeSIMLA is an effective platform for simulating large-scale genetic data. Each individual pool was expanded to 100,000 chromosomes before termination, which typically took less than 10 minutes including LD calculation. Additionally, methods other than purely stochastic independent initialization for pools of chromosomes could be used, which could lead to superior data properties and less generations of population growth.

The speed and scale of the genomeSIMLA software is sufficient to provide timely results to investigators conducting power studies for various methods. The software architecture ensures that the user can access all available computational power to do very large whole-genome size studies. The time to completion for various size simulations for single and multiple processors are presented in Table 2. Those times include the time required to calculate and provide an interactive graphical interface of LD pictures for locus selection. These times are very fast given the computational task and represent the advanced implementation which is presented here. Demonstrations, manuals, and genomeSIMLA software for Mac, PC, and Linux are available for download at http://chgr.mc.vanderbilt.edu/genomeSIMLA. With this capability, researchers who develop novel methods to analyze genetic data can quickly and accurately estimate the performance and sampling properties of those methods.

Acknowledgements

This work was supported by National Institutes of Health grants HL65962 and AG20135.

References

1. The International HapMap Project. Nature, 426, 789-796 (2003)
2. Dudek, S., Motsinger, A.A., Velez, D., Williams, S.M., Ritchie, M.D.: Data simulation software for whole-genome association and other studies in human genetics. In: Pac Symp Biocomput, pp. 499–510 (2006)
3. Barrett, J.C., Cardon, L.R.: Evaluating coverage of genome-wide association studies. Nat Genet 38, 659–662 (2006)
4. Hunter, D.J., Kraft, P., Jacobs, K.B., Cox, D.G., Yeager, M., Hankinson, S.E., et al.: A genome-wide association study identifies alleles in FGFR2 associated with risk of sporadic postmenopausal breast cancer. Nat Genet, Jul 1939, 870–874 (2007)
5. Easton, D.F., Pooley, K.A., Dunning, A.M., Pharoah, P.D., Thompson, D., Ballinger, D.G., et al.: Genome-wide association study identifies novel breast cancer susceptibility loci. Nature 447, 1087–1093 (2007)
6. Genome-wide association study of 14,000 cases of seven common diseases and 3,000 shared controls. Nature 447, 661–678 (2007)
7. Saxena, R., Voight, B.F., Lyssenko, V., Burtt, N.P., de Bakker, P.I., Chen, H., et al.: Genome-wide association analysis identifies loci for type 2 diabetes and triglyceride levels. Science 316, 1331–1336 (2007)
8. Scott, L.J., Mohlke, K.L., Bonnycastle, L.L., Willer, C.J., Li, Y., Duren, W.L., et al.: A genome-wide association study of type 2 diabetes in Finns detects multiple susceptibility variants. Science 316, 1341–1345 (2007)
9. Zeggini, E., Weedon, M.N., Lindgren, C.M., Frayling, T.M., Elliott, K.S., Lango, H., et al.: Replication of genome-wide association signals in UK samples reveals risk loci for type 2 diabetes. Science 316, 1336–1341 (2007)
10. Lyon, H.N., Emilsson, V., Hinney, A., Heid, I.M., Lasky-Su, J., Zhu, X., et al.: The association of a SNP upstream of INSIG2 with body mass index is reproduced in several but not all cohorts. PLoS Genet, e61 (2007)
11. Schmidt, M., Hauser, E.R., Martin, E.R., Schmidt, S.: Extension of the SIMLA package for generating pedigrees with complex inheritance patterns: environmental covariates, gene-gene and gene-environment interaction. Stat Appl Genet Mol Biol, 2005, Article15 (2004)
12. Bass, M.P., Martin, E.R., Hauser, E.R.: Pedigree generation for analysis of genetic linkage and association. In: Pac Symp Biocomput, pp. 93–103 (1993)
13. Marchini, J., Donnelly, P., Cardon, L.R.: Genome-wide strategies for detecting multiple loci that influence complex diseases. Nat Genet 37, 413–417 (2005)
14. Boehnke, M.: Estimating the power of a proposed linkage study: a practical computer simulation approach. Am J Hum Genet 39, 513–527 (1986)
15. Ploughman, L.M., Boehnke, M.: Estimating the power of a proposed linkage study for a complex genetic trait. Am J Hum Genet 44, 543–551 (1989)
16. Weeks, D.E., Ott, J., Lathrop, G.M.: SLINK: A general simulation paorgram for linkage analysis. American Journal of Human Genetics 47, A204 (1990)
17. Kingman, J.: The coalescent. Stochastic Processes Appl 13, 235–248 (1982)

18. Liang, L., Zollner, S., Abecasis, G.R.: GENOME: A rapid coalescent-based whole genome simulator. Bioinformatics 23, 1565–1567 (2007)
19. Wright, F.A., Huang, H., Guan, X., Gamiel, K., Jeffries, C., Barry, W.T., et al.: Simulating association studies: A data-based resampling method for candidate regions or whole genome scans. Bioinformatics 23, 2581–2588 (2007)
20. Balloux, F.: EASYPOP (version 1.7): A computer program for population genetics simulations. J Hered 92, 301–302 (2001)
21. Hey, J.: A computer program for forward population genetic simulation, Ref Type: Computer Program (2005)
22. Hoggart, C.J., Chadeau, M., Clark, T.G., Lampariello, R., De, I.M., Whittaker, J.C., et al.: Sequence-level population simulations over large genomic regions. Genetics (2007)
23. Peng, B., Kimmel, M.: simuPOP: A forward-time population genetics simulation environment. Bioinformatics (2005)
24. Moore, J.H., Hahn, L.W., Ritchie, M.D., Thornton, T.A., White, B.: Routine Discovery of High-Order Epistasis Models for Computational Studies in Human Genetics. Applied Soft Computing 4, 79–86 (2004)
25. Richards, F.: A flexible growth function for empirical use. Journal of Experimental Botany 10, 290–300 (1959)
26. Durrant, C., Zondervan, K.T., Cardon, L.R., Hunt, S., Deloukas, P., Morris, A.P.: Linkage disequilibrium mapping via cladistic analysis of single-nucleotide polymorphism haplotypes. Am J Hum Genet 75, 35–43 (2004)
27. Excoffier, L., Slatkin, M.: Maximum-likelihood estimation of molecular haplotype frequencies in a diploid population. Mol Biol Evol 12, 921–927 (1995)
28. Stephens, M., Scheet, P.: Accounting for decay of linkage disequilibrium in haplotype inference and missing-data imputation. Am J Hum Genet 2005, 449-462 (March, 1976)

DEEPER: A Full Parsing Based Approach to Protein Relation Extraction

Timur Fayruzov[1], Martine De Cock[1], Chris Cornelis[1], and Véronique Hoste[2]

[1] Department of Applied Mathematics and Computer Science
Ghent University Association
Krijgslaan 281 (S9), 9000 Gent, Belgium
{Timur.Fayruzov,Martine.DeCock,Chris.Cornelis}@UGent.be
[2] School of Translation Studies
Ghent University Association
Groot-Brittanniëlaan 45, 9000 Gent, Belgium
Veronique.Hoste@hogent.be

Abstract. Lexical variance in biomedical texts poses a challenge to automatic protein relation mining. We therefore propose a new approach that relies only on more general language structures such as parsing and dependency information for the construction of feature vectors that can be used by standard machine learning algorithms in deciding whether a sentence describes a protein interaction or not. As our approach is not dependent on the use of specific interaction keywords, it is applicable to heterogeneous corpora. Evaluation on benchmark datasets shows that our method is competitive with existing state-of-the-art algorithms for the extraction of protein interactions.

1   Introduction

Studying the interactions of proteins is an essential task in biomedical research, so it comes as no surprise that a lot of effort is being devoted to the construction of protein interaction knowledge bases. More and more relevant information is becoming available on the web, in particular in literature databases such as MEDLINE[1], in ontological resources such as the Gene Ontology[2], and in specialized structured databases such as IntAct[3]. The unstructured information in scientific publications poses the biggest challenge to biologists who are interested in specific gene or protein interactions, as they are forced to spend a tremendous amount of time reviewing articles looking for the information they need. Structured knowledge bases are easier to query, but again require a great deal of knowledge and labour intensive maintenance to stay synchronized with the latest research findings in molecular biology. Automation tools can facilitate this task, which is why machine learning techniques for information extraction (IE) in the biomedical domain have gained a lot of attention over the last years.

---

[1] http://www.ncbi.nlm.nih.gov/
[2] http://www.geneontology.org/
[3] http://www.ebi.ac.uk/intact/site/index.jsf

E. Marchiori and J.H. Moore (Eds.): EvoBIO 2008, LNCS 4973, pp. 36–47, 2008.

Relation extraction from texts is one of the most difficult tasks of IE. In natural language, relations can be expressed in different ways, hence no universal set of rules or patterns for mining them can be constructed. Traditional algorithms for relation learning from texts can perform reasonably well (see e.g. [1,2,5,12]), but they typically rely explicitly or implicitly on specific interaction keywords, which limits their applicability to heterogeneous data. The biggest obstacle with heterogeneous datasets is that they describe protein interactions using different lexicons. However, entirely different surface representations for interactions can still exhibit the same syntactic pattern. We therefore propose to abstract from pure lexical data and to concentrate only on more general language structures such as parsing and dependency information. This coarser grained approach allows to cope better with the lexical variance in the data. Indeed, taking the fact into account that lexically different expressions of protein interactions might still bear some resemblance on the syntactic level provides welcome hints for machine learning techniques that commonly thrive on similarities in the data.

The resulting system is a mining tool that facilitates information extraction and knowledge base maintenance by presenting to the user protein interactions identified in scientific texts. The tool aims at supporting biologists in finding relevant information, rather than to exclude them entirely from the data processing flow. After reviewing related approaches in Section 2, we give a detailed description of the proposed method in Section 3. Abstracting from pure lexical data and only relying on syntactic patterns instead bears the risk of overgeneralizing, in the sense that sentences that do not describe protein interactions might exhibit a syntactic structure similar to those that do, and hence they might get incorrectly identified as protein interactions. To verify the reliability of our approach we therefore evaluated it on two benchmark datasets. The experimental results and a comparison with existing algorithms are described in Section 4. Concluding remarks and future work are presented in Section 5.

2   Related Work

The extraction of protein relations has attracted a lot of attention during the last years, resulting in a range of different approaches. The first step is the recognition of the protein names themselves (see e.g. [3,6,15]). As the focus of this paper is on the mining of *interactions*, we assume that protein name recognition has already taken place. The recognition of protein interactions is typically treated as a classification problem: the classifier gets as input information about a sentence containing two protein names and decides whether the sentence describes an actual interaction between those proteins or not. The classifier itself is built manually or, alternatively, it is constructed automatically using an annotated corpus as training data. The different approaches can be distinguished based on the information they feed to the classifier: some methods use only shallow parsing information on the sentence while others exploit full parsing information.

Shallow parsing information includes part-of-speech (POS) tags and lexical information such as lemmas (the base form of words occuring in the sentence) and orthographic features (capitalization, punctuation, numerals etc.). In [2], a support vector machine (SVM) model is used to discover protein interactions. In this approach each sentence is split into three parts — before the first protein, between the two proteins and after the second protein. The kernel function between two sentences is computed based on common sequences of words and POS tags. In another approach [5], this kernel function is modified to treat the same parts of the sentence as bags-of-words and called a global context kernel. It is combined with another kernel function called a local context kernel, that represents a window of limited size around the protein names and considers POS, lemmas, and orthographic features as well as the order of words. The resulting kernel function in this case is a linear combination of the global context kernel, and the local context kernel.

A completely different approach is presented in [12], where very high recall and precision rates are obtained by means of hand-crafted rules for sentence splitting and protein relation detection. The rules are based on POS and keyword information, and they were built and evaluated specifically for *Escherichia coli* and yeast domains. It is questionable, however, whether comparable results could be achieved in different biological domains and how much effort would be needed to adapt the approach to a new domain. In another approach reported on in [8], a system was built specifically for the LLL challenge (see Section 4). First, training set patterns are built by means of pairwise sentence alignment using POS tags. Next, a genetic algorithm is applied to build several finite state automata (FSA) that capture the relational information from the training set.

Besides the methods described above, approaches have been proposed that augment shallow parsing information with full parsing information, i.e. syntactic information such as full parse and dependency trees. In [4] for instance, for every sentence containing two protein names a feature vector is built containing terms that occur in the path between the proteins in the dependency tree of the sentence. These feature vectors are used to train an SVM based classifier with a linear kernel. More complex feature vectors are used in [10], where the local contexts of the protein names, the root verbs of the sentence, and the parent of the protein nodes in the dependency tree are taken into account by a BayesNet classifier. In [7], syntactic information preprocessing, hand-made rules, and a domain vocabulary are used to extract gene interactions. The approach in [17] uses predicate-argument structures (PAS) built from dependency trees. As surface variations may exhibit the same PAS, the approach aims at tackling lexical variance in the data. It is tailored towards the AImed dataset (see Section 4) for which 5 classes of relation expression templates are predefined manually. The classes are automatically populated with examples of PAS patterns and protein interactions are identified by matching them against these patterns.

To the best of our knowledge, all existing work either uses only shallow parsing information (including lexical information) or a combination of shallow and full parsing information. Furthermore, approaches of the latter kind typically use dependency trees only as a means to e.g. detect chunks or to extract relevant keywords. The goal of this paper is to investigate what can be achieved using *only* full parsing information. In other words, the full parsing information is not used as a means to select which further (lexical) information to feed to the classifier, but it is used as a direct input itself to the classifier. The fact that such an approach is independent of the use of a specific lexicon makes it worthwhile to investigate.

3   DEEPER: A Dependency and Parse Tree Based Relation Extractor

There is an abundance of ways in English to express that proteins stimulate or inhibit one another, and the available annotated corpora on protein interactions cover only a small part of them. In other words, when the interaction mining tool is confronted with a previously unseen text, it is likely for this text to contain protein interactions described in ways for which there is no exact matching example in the training data. However, different surface representations can still exhibit a similar syntactic pattern, as the following example illustrates.

Example 1. Consider the following sentences about the interaction between *sigma F* and *sigma E* in one case and between *GerE* and *cotB* in the other case:

Sigma F activity regulates the processing of sigma E within the mother cell compartment.

A low GerE concentration, which was observed during the experiment, activated the transcription of cotB by final sigmaK RNA polymerase, whereas a higher concentration was needed to activate transcription of cotX or cotC.

Although the surface representations are very different, the underlying syntactic pattern, which represents part of a dependency tree, is the same in both cases:

$$protein \xleftarrow{nn} noun \xleftarrow{nsubj} verb \xrightarrow{dobj} noun \xrightarrow{prep\_of} protein$$

We exploit this deeper similarity between training instances by using dependency and parsing information to build abstract representations of interactions. Such representations have less variance than the initial lexical data, hence sensible results can be obtained from smaller training datasets. The approach is fully automatical and consists of three stages: after a text preprocessing stage, for every sentence containing two tagged protein names, we construct a feature vector summarizing relevant information on the parse tree and the dependency tree. In the third stage a classifier is trained to recognize whether the sentence describes an actual interaction between the proteins. The novelty of the approach w.r.t. existing work is that we do not use dependency data to detect keywords, but we consider dependencies as features themselves. In the next section we show that using only this kind of syntactic information without any lexical data allows to obtain reasonably good results.

Text Preprocessing. This step is intended to simplify the sentence structure and hence increase the parser reliability. It includes sentence splitting as well as the detection and the substitution of complex utterances (e.g. chemical formulas or constructions with many parentheses) with artificial strings, which in some cases can otherwise significantly reduce the quality of parsing. Furthermore, we expand repeating structures, turning e.g. 'sigA- and sigB-proteins' or 'sigA/ sigB-proteins' into 'sigA-proteins and sigB-proteins'. All substitutions are done automatically by means of regular expressions; hence the same kind of preprocessing can be applied to an arbitrary text. Moreover, tagged protein names in the text may include more than one word; in order to treat them as a single entity in further processing stages, we replace them in the same manner as formulas. Finally, we take all possible pairwise combinations of proteins in each sentence and create one sentence for each combination where only this combination is tagged. Part of this process is shown in Example 2.

Example 2. Below is the result after text preprocessing for the second sentence from Example 1:

> A low <u>GerE</u> concentration, which was observed during the experiment, activated the transcription of <u>cotB</u> by final sigmaK RNA polymerase, whereas a higher concentration was needed to activate transcription of cotX or cotC.
> . . .
> A low GerE concentration, which was observed during the experiment, activated the transcription of cotB by final sigmaK RNA polymerase, whereas a higher concentration was needed to activate transcription of <u>cotX</u> or <u>cotC</u>.

Feature Vector Construction. After the text preprocessing stage, for each sentence we build a feature vector that summarizes important syntactic information on the parse tree and the typed dependency tree, which are both ways of representing sentence structure. A parse tree is a tree (in terms of graph theory) that represents the syntactical structure of a sentence. Words from the sentence are leaves of the parse tree and syntactical roles are intermediate nodes, so a parse tree represents the nesting structure of multi-word constituents. A dependency tree on the other hand represents interconnections between individual words of the sentence. Hence, all nodes of the dependency tree are words of the sentence, and edges between nodes represent syntactic dependencies. In typed dependency trees, edges are labeled with syntactic functions (e.g., subj, obj). Figure 1 depicts the typed dependency tree and parse tree for the first sentence of Example 1.

During the feature extraction phase we parse each sentence with the Stanford Parser[4]. For each tagged pair of proteins (recall that each sentence has only one such pair), we extract a linked chain [14] from the dependency tree. The dependency tree is unordered w.r.t. the order of the words in the sentence; hence to produce patterns uniformly, we order the branches in the linked chain based

---

[4] http://nlp.stanford.edu/downloads/lex-parser.shtml

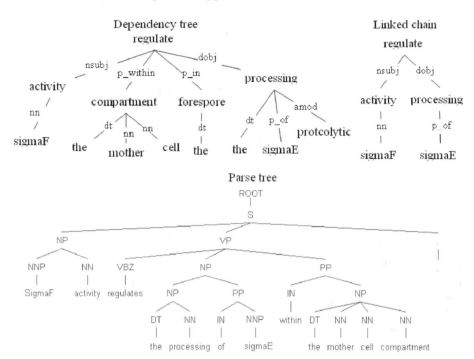

Fig. 1. Dependency and parse trees and linked chain for the first sentence of Ex. 1

on the position of the words in the initial sentence. Thus the left branch contains the word that occurs earlier in the sentence and the right branch the word that occurs later. The absolute majority of the branches in the linked chains from the datasets we examined contain no more than 6 edges, and those which contain more are negative instances, so we choose feature vectors with 6 features for each branch to cover all positive examples from the training set. Therefore, we use the first 6 dependencies from the left branch as the first 6 features in the feature vector. Likewise, the first 6 dependencies from the right branch correspond to features 7 through 12. Moreover, to better describe the structure of the relation we incorporate information from the parse tree as well, namely the length of the path from the root of the parse tree to each protein as the 13th and the 14th feature, and the number of nested subsentences in these paths as the 15th and the 16th feature.

Example 3. Below is the feature vector of the first sentence from Example 1:

$nsubj$	nn				$dobj$	$prep\_of$				4	7	0	0

We extract a linked chain between the two proteins, as shown in Figure 1. It is already ordered, i.e. *Sigma F* precedes *Sigma E* in the sentence, so we do not need to swap these branches. We take the left branch and fill the first 6 features of the feature vector. As the branch contains only 2 dependencies — *nsubj* and *nn*, 4 slots in the vector remain empty. Features 7-12 for the right branch are filled in the same manner. *Sigma F* is at depth 4 in the parse tree while *Sigma E* is at depth 7, and the parse tree in Figure 1 does not contain subsentences. All this information is reflected in the last 4 features of the vector. Note that the resulting feature vector contains several empty fields; only the most complicated sentences will have a value for each feature in the vector.

Training a Classifier. By the above process, we obtain a set of feature vectors for sentences which can be divided into two classes — those that describe real protein interactions and those that do not. Therefore, we can use a standard classification algorithm to distinguish between these two classes. To build the classifier, we use a decision tree algorithm (C4.5 implementation [13]) and the BayesNet classifier [9]. These two algorithms represent classical instances of two branches of machine learning — rule induction and statistical learning — which employ different approaches to data processing. Decision trees consist of internal nodes which represent conditions on feature values, and leaves which represent classification decisions that conform to the feature values in nodes on the way to this leaf. The BayesNet classifier is represented as a directed graph with a probability distribution for each feature in the nodes and with the edges denoting conditional dependencies between different features. When we use the BayesNet classifier we apply a conditional independence assumption, which means that probabilities of node values depend only on probabilities of values of their immediate parents, and do not depend on higher ancestors. This corresponds to the reasonable assumption that the syntactic role of a node in the linked chain depends on the syntactic role of its parent only.

To overcome the problem of missing values (which occur frequently in the feature vectors), in the BayesNet classifier we simply change them by a default value. With C4.5, to classify an instance that have a missing value for a given node, the instance is weighted proportionally to the number of instances that go down to each branch, and recursively processed on each child node w.r.t. to assigned weight. This process is described in more detail in [16].

4   Experimental Evaluation

To verify the reliability of our approach, we evaluated it on two datasets. The first dataset [11] originates from the Learning Language in Logic (LLL) relation mining challenge on Genic Interaction Extraction[5]. This dataset contains annotated protein/gene interactions concerned with *Basilicus subtilis* transcription. The sentences in the dataset do not make up a full text, but they are individual sentences taken from several abstracts retrieved from Medline. The proteins involved in the interactions are annotated with agent and target roles; because our current approach is not aimed at mining the direction of interactions, we ignore this annotated information and treat the interactions as symmetrical relations.

---

[5] http://genome.jouy.inra.fr/texte/LLLchallenge/

Table 1. Datasets used in the experiments

Dataset	# sentences	# positive instances	# negative instances
AImed	1978	816	3204
LLL'05	77	152	233

The AImed dataset [1] is compiled from 197 abstracts extracted from the Database of Interacting Proteins (DIP) and 28 abstracts which contain protein names but do not contain interactions. Since we are interested in retrieving protein interactions, in this paper we use only the former set of abstracts. The connection between a full name of a protein and its abbreviation, e.g. *tumor necrosis factor (TNF)*, is annotated as an interaction in the AImed dataset. Since such an annotation is not concerned with an actual interaction between different proteins, we omit this kind of data from our experiments. Furthermore we removed nested protein annotations, which wrap around another protein or interaction annotation. Finally, TI- and AD- sections as well as PG- prefixes, which are Medline artifacts, were removed.

More information about the datasets is listed in Table 1. From this table, it is clear that the AImed dataset is highly imbalanced, as there is a strong bias to negative examples. To the best of our knowledge, these are the only two publicly available datasets containing annotations of protein interactions and hence suitable to evaluate our approach. In the evaluation we used 10-fold cross validation for both the AImed and the LLL05 dataset; furthermore we ran experiments with AImed as training set and LLL05 as test set. We used Weka [16] for the implementation of the machine learning methods.

The difference in the datasets requires different parameters to achieve optimal performance. As we have mentioned above, the AImed dataset is imbalanced and using it for training tends to lead to a bias towards classifying examples as negative (independently of the training scheme). For this reason, we use cost-sensitive learning [16] to decrease the bias when AImed is used as a training set. Moreover, in the C4.5 implementation for the AImed dataset, we build a binary decision tree, i.e. at each node the algorithm tests only one value of one feature. Otherwise, the algorithm would decide that the empty tree classifies the dataset in the best way, and all examples would be classified as negative (again, because of the biased dataset).

The results below are described in terms of the sentences that are (in)correctly identified by the system as describing a protein interaction, as these are exactly the instances that the system will present to the biologist. The *relevant instances* are the sentences that should have been identified as describing protein interactions; this includes the true positives, i.e. the positive instances that are correctly identified by the system, but also the false negatives, i.e. the positive instances that are overlooked by the system. The *retrieved instances* are the sentences that are identified by the system as describing protein reactions. This includes the true positives but may also include false positives, i.e. sentences incorrectly identified by the system as describing a protein interaction. Using TP, FN, and

FP to denote the number of true positives, false negatives, and false positives respectively, recall and precision are defined as

$$\text{recall} = \frac{\text{TP}}{\text{TP} + \text{FN}} \qquad \text{precision} = \frac{\text{TP}}{\text{TP} + \text{FP}}$$

Recall (also referred to as coverage) indicates how many of the relevant instances are retrieved. Precision (also referred to as accuracy) indicates how many of the retrieved instances are relevant.

To study the trade-off between recall and precision we use a confidence threshold p between 0 and 1 such that an instance is retrieved iff the classifier has a confidence of at least p that the instance describes a real protein interaction. The BayesNet classifier provides such a confidence value naturally, because its output is a class distribution probability for each instance. Decision trees can also be easily adapted to produce a probabilistic output by counting training examples at the leaf nodes. If a sentence that is being classified ends up at a leaf node, the confidence of the classifier that it is a positive instance, is the proportion of positive training examples to all training examples at that leaf node. When p is set to 1, a sentence is only retrieved if the classifier has absolute confidence that it describes a protein interaction. In this case typically the precision is high while the recall is low. Decreasing the threshold p allows to increase the recall at the cost of a drop in the precision. Figure 2 shows recall-precision curves obtained by varying p from 1 to 0.

As the first picture depicts, both classifiers allow to obtain similarly nice results for the LLL05 dataset, which is a first indication that we can make reasonable predictions about the occurrence of protein interactions in sentences based solely on full parsing information. Several authors present results of their protein relation extraction methods on the LLL05 dataset. However, since our current approach is not aimed at identifying agent and target roles in the interactions, we can only compare our results with those methods that treat the interactions as symmetrical relations. The first picture in Figure 2 shows a result from [7] depicted by a * and corresponding to a recall of 85% and a precision of 79%. One should keep in mind that the method from [7] uses hand-made rules and a domain vocabulary, while our approach does not employ any prespecified knowledge. However, results show that our approach with a C4.5 classifier achieves results which are very close to the ones obtained by RelEx.

Whereas the LLL05 dataset contains only selected sentences from Medline abstracts, the AImed dataset contains full abstracts, posing a bigger challenge to our approach. The second picture in Figure 2 shows that C4.5 and BayesNet allow to obtain comparable results in terms of recall and precision. They both outperform the PAS-approach for which a recall of 33.1% for a precision of 33.7% is reported in [17].

Finally, we performed a cross dataset experiment using the AImed dataset for training the classifier and the LLL05 dataset for testing. The corresponding recall-precision curves for C4.5 and the BayesNet classifier are shown in the third picture in Figure 2. While both datasets are independent (built for different biological subdomains and by different people), our approach shows good results.

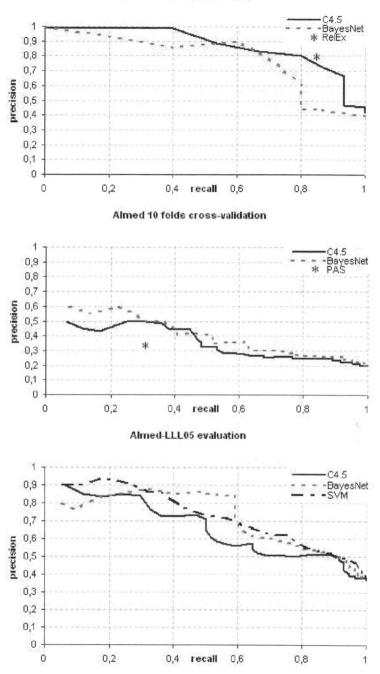

Fig. 2. Recall-precision charts

This indicates that the current approach is applicable to different domains without alterations, although further investigation is needed to back up this claim. The third picture also shows the recall-precision curve for the subsequence kernel method from [2] which is a state-of-the-art shallow parsing based approach for relation extraction. Since the approach was evaluated on a different dataset in [2], we used the implementation provided by the authors[6] and our datasets to perform the experiment. The training is done with LibSVM[7] on the AImed dataset and testing is done on the LLL05 dataset. The results show that the three methods are comparable, with a slight preference for our approach with the BayesNet classifier, as it can keep up a very high precision of 84% for a recall of up to 60%.

5   Conclusions and Future Work

Whereas existing approaches for protein interaction detection typically rely on shallow parsing information, sometimes augmented with full parsing information, we presented an approach based solely on full parsing information. More in particular, we proposed a clean and generally applicable approach in which for each sentence a feature vector is constructed that contains 12 features with information on the dependency tree and 4 features with information on the parse tree of the sentence. Next we fed these feature vectors as inputs to a C4.5 and a BayesNet classifier, as representatives of a rule induction and a statistical learning algorithm. Using these standard data mining algorithms and no shallow parsing or lexical information whatsoever, we were able to obtain results which are comparable with state-of-the-art approaches for protein relation mining. This result is promising since a method that uses only full parsing information does not depend on specific interaction keywords and is less affected by the size and/or the heterogenity of the training corpus.

As this paper presents work in progress, quite some ground remains to be covered, including a more complete comparison with existing methods. Among other things, it would be interesting to build an SVM model with our feature vectors and compare the results with those of shallow and combined parsing based approaches that rely on kernel methods as well. Furthermore, we intend to look into detecting the agents and the targets of interactions, which would allow us to do an independent evaluation on the LLL05 dataset as intended by the LLL challenge. A final intriguing question is whether an augmentation with shallow parsing information could increase the performance of our approach.

Acknowledgement

Chris Cornelis would like to thank the Research Foundation–Flanders for funding his research.

---

[6] http://www.cs.utexas.edu/~razvan/code/ssk.tar.gz
[7] http://www.csie.ntu.edu.tw/~cjlin/libsvm/

References

1. Bunescu, R., Ge, R., Kate, J.R., Marcotte, M.E., Mooney, R.J., Ramani, K.A., Wong, W.Y.: Comparative Experiments on Learning Information Extractors for Proteins and their Interactions. Artificial Intelligence in Medicine 33(2) (2005)
2. Bunescu, R., Mooney, J.R.: Subsequence Kernels for Relation Extraction. In: Proc. 19th Conf. on Neural Information Processing Systems (NIPS) (2005)
3. Collier, N., Nobata, C., Tsujii, J.: Extracting the Names of Genes and Gene Products with a Hidden Markov Model. Iin: Proc. 17th Int. Conf. on Computational Linguistics (2000)
4. Erkan, G., Ozgur, A., Radev, D.R.: Extracting Interacting Protein Pairs and Evidence Sentences by using Dependency Parsing and Machine Learning Techniques. In: Proc. 2nd BioCreAtIvE Challenge Workshop — Critical Assessment of Information Extraction in Molecular Biology (2007)
5. Giuliano, C., Lavelli, A., Romano, L.: Exploiting Shallow Linguistic Information for Relation Extraction From Biomedical Literature. Iin: Proc. 11th Conf. of the European Chapter of the Association for Computational Linguistics (EACL (2006)
6. Fukuda, K., Tamura, A., Tsunoda, T., Takagi, T.: Toward Information Extraction: Identifying Protein Names from Biomedical Papers. In: Proc. Pacific Symp. on Biocomputing (1998)
7. Fundel, K., Küffner, R., Zimmer, K.: RelEx—Relation extraction using dependency parse trees, Bioinformatics 23(3) (2007)
8. Hakenberg, J., Plake, C., Leser, U., Kirsch, H., Rebholz-Schuhmann, D.: LLL 2005 Challenge: Genic Interaction Extraction — Identification of Language Patterns Based on Alignment and Finite State Automata. In: Proc. ICML 2005 Workshop: Learning Language in Logic (2005)
9. Jensen, F.V.: An Introduction to Bayesian Networks. Springer, Heidelberg (1996)
10. Katrenko, S., Adriaans, P.: Learning Relations from Biomedical Corpora Using Dependency Tree Levels. In: Proc. BENELEARN conference (2006)
11. Nédellec, C.: Learning Language in Logic - Genic Interaction Extraction Challenge. In: Proc. ICML 2005 Workshop: Learning Language in Logic
12. Ono, T., Hishigaki, H., Tanigami, A., Takagi, T.: Automated Extraction of Information on Protein-Protein Interactions from the Biological Literature. Bioinformatics, 17(2) (2001)
13. Quinlan, J.R.: C4.5: Programs for Machine Learning. Morgan Kaufmann, San Francisco (1993)
14. Stevenson, M., Greenwood, M.A.: Comparing Information Extraction Pattern Models. In: Proc. IE Beyond The Document Workshop (COLING/ACL (2006)
15. Tanabe, L., Wilbur, W.J.: Tagging Gene and Protein Names in Biomedical Text. Bioinformatics, 18(8) (2002)
16. Witten, I.H., Frank, E.: Data Mining: Practical Machine Learning Tools and Techniques, 2nd edn. Morgan Kaufmann, San Francisco (2005)
17. Yakushiji, A., Miyao, Y., Tateisi, Y., Tsujii, J.: Biomedical Information Extraction with Predicate-Argument Structure Patterns. In: Proc. 1st Int. Symp. on Semantic Mining in Biomedicine (2005)

Improving the Performance of Hierarchical Classification with Swarm Intelligence

Nicholas Holden and Alex A. Freitas

Computing Laboratory, University of Kent,
Canterbury, CT2 7NF, UK
nickpholden@gmail.com, A.A.Freitas@kent.ac.uk

Abstract. In this paper we propose a new method to improve the performance of hierarchical classification. We use a swarm intelligence algorithm to select the type of classification algorithm to be used at each "classifier node" in a classifier tree. These classifier nodes are used in a top-down divide and conquer fashion to classify the examples from hierarchical data sets. In this paper we propose a swarm intelligence based approach which attempts to mitigate a major drawback with a recently proposed local search-based, greedy algorithm. Our swarm intelligence based approach is able to take into account classifier interactions whereas the greedy algorithm is not. We evaluate our proposed method against the greedy method in four challenging bioinformatics data sets and find that, overall, there is a significant increase in performance.

Keywords: Particle Swarm Optimisation, Ant Colony Optimisation, Data Mining, Hierarchical Classification, Protein Function Prediction.

1 Introduction

Hierarchical classification is a challenging area of data mining. In hierarchical classification the classes are arranged in a hierarchical structure, typically a tree or a DAG (Directed Acyclic Graph). In this paper we consider classes arranged in a tree structure where each node (class) has only one parent – with the exception of the root of the tree, which does not have any parent and does not correspond to any class. Hierarchical class datasets present two main new challenges when compared to flat class datasets. Firstly, many (depending on the class depth) more classes must be assigned to the examples. Secondly, the prediction of a class becomes increasingly difficult as deeper class levels are considered, due to the smaller number of examples per class.

In this paper we address the problem of hierarchical protein function prediction, a very active research topic in bioinformatics. The prediction of protein function is one of the most important challenges faced by biologists in the current "post-genome" era. The challenge lies in the fact that the number of proteins discovered each year is growing at a near exponential rate [1] (with the vast majority of them having unknown function) and advances in the understanding of protein function are critical for more effective diagnosis and treatment of disease, also helping in the design of more effective medical drugs etc.

E. Marchiori and J.H. Moore (Eds.): EvoBIO 2008, LNCS 4973, pp. 48–60, 2008.

In this paper we propose a new method to increase the accuracy of classification when using the top-down divide and conquer (TDDC) approach for hierarchical classification (as described in section 2). The new method is based on a swarm intelligence algorithm, more precisely a hybrid particle swarm optimisation/ant colony optimisation (PSO/ACO) algorithm.

The remainder of this paper is organised as follows: Section 2 introduces hierarchical classification. Section 3 describes an approach proposed by Secker at al. [3] for improving hierarchical classification accuracy and critiques it. Section 4 describes the proposed novel method for hierarchical classification using a swarm intelligence (PSO/ACO) algorithm. Section 5 describes experimental set-up. Section 6 describes the experimental data from four challenging "real-world" biological data sets and section 7 draws conclusions based on the results of the experiments and suggests future research directions.

2   A Brief Review of Hierarchical Classification

This paper focuses on hierarchical classification problems where the classes to be predicted are organized in the form of a tree, hereafter referred to as a class tree populated by class nodes. An example of a hierarchical classification problem might be the prediction of what species and then breed a pet is. In the first case we wish to known whether the given animal is of the class node (species) dog or cat, and in the second case if the animal is of the class node (breed) Burmese, British Blue, Jack Russell or Golden Retriever. In this paper the species would be considered the first class level and the breed the second class level. The TDDC approach is based on the principle that only sibling class nodes need be considered at any point in the hierarchical tree. So at the first set of sibling class nodes (cat or dog) if we decide cat, then at the second set of class nodes we must only decide between the sibling class nodes Burmese or British Blue. Notice that this has a major drawback, which is that if the pet is in fact a dog we are guaranteed to guess the breed wrong if we predict cat at the first class level.

Fig. 1. A Hierarchical classification problem using the TDDC approach

This top-down approach has the important advantage of using information associated with higher-level classes in order to guide the prediction of lower-level classes. This has shown to increase accuracy over other basic approaches [4]. For instance, (from Figure 1), if class 1.X (where X denotes any digit) is predicted at the first level and that class node only has the child nodes 1.1 and 1.2, only these two class nodes should be considered and not the children belonging to node 2.X, 2.1 and 2.2. In Figure 1 the classifier nodes are shown by the grey boxes. There would be classifiers to distinguish between classes 1.X and 2.X, 1.1 and 1.2 etc. It is important to distinguish between two conceptually distinct – though clearly related – trees, namely a class tree and a classification tree. In the class tree every node represents a class (to be predicted by a classifier). By contrast, in the TDDC tree each node represents a classifier that discriminates between sibling classes. The nodes of a classifier tree are hereafter referred to as classifier nodes. The terms classifier tree and TDDC tree are used interchangeably in this paper.

3   The Greedy Selective Top Down Divide and Conquer Approach

In the conventional top-down approach for hierarchical classification, in general, the same classification algorithm is used for each classifier node. Intuitively, this is a suboptimal approach because each classifier node is associated with a different classification problem – more precisely, a different training set, associated with a different set of classes to be predicted. This suggests that the predictive accuracy of the classifier tree can be improved by selecting, at each classifier node, the classification algorithm with best performance in the classification problem associated with each node, out of a predefined list of candidate classification algorithms. Indeed it was found in [3] by Secker et al. that by varying the classification algorithm at each classifier node in the Top-Down Divide and Conquer (TDDC) tree, classification accuracy could, in general, be somewhat improved.

In Secker's work the training set at each classifier node is divided into two non overlapping sub sets, a building set – used to train the classification algorithms – and a separate validation set – which is used to assess the predictive accuracy of the models constructed by the classification algorithms. At every classifier node in the TDDC tree, multiple classifiers are built using the building set, each using a different classification algorithm. The classification accuracy of each of these classifiers is measured using the validation set at each classifier node, and then the best classifier (according to classification accuracy in the validation set) is chosen. This process is repeated at each classifier node to select a set of classifiers to populate the TDDC classification tree, which is then used to classify the test instances (unseen during training). A simple example of a classification tree constructed by this method, showing a different classifier chosen at each node, is shown in Figure 2.

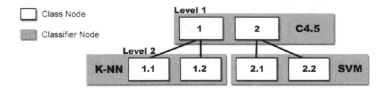

Fig. 2. A TDDC tree using classification algorithm selection

In this way Secker's work uses a greedy selective approach to try and maximise classification accuracy. It is described as greedy because, when it selects a classifier at each classifier node, it maximises accuracy only in the current classifier node, using local data. Therefore, the greedy selective approach ignores the effect of this local selection of a classifier on the entire classifier tree. In other words, this procedure is "short sighted", and so it does not consider the interaction between classifiers at different classifier nodes.

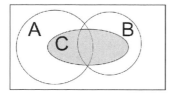

Fig. 3. Classifier interaction scenario where $|B \cap C| > |A \cap C|$

Fig. 4. Classifier interaction scenario where $|B \cap C| < |A \cap C|$

Figures 3 and 4 show two possible scenarios demonstrating interactions between classifiers at different classifier nodes during classifier evaluation. A and B are the two possible parent classifiers trying to discriminate between classes 1 and 2. C is the child classifier that attempts to discriminate between classes 1.1 and 1.2 – as shown in Figure 5. Figures 3 and 4 show the sets of correctly classified examples for each classifier in the TDDC tree. Notice that $C \subseteq A \cup B$ for the three classifiers A, B and C. This is due to the fact that in the standard TDDC tree once a misclassification has been made, by classifiers A or B at the first classifier node, it cannot be rectified by C at the child classifier node.

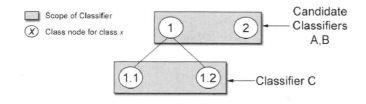

Fig. 5. A class tree used to illustrate the discussion on classifier interaction

As mentioned earlier, the greedy approach chooses the best classifier at each node according to the classification accuracy, in the validation set, at that node. In the scenarios shown in both Figures 3 and 4 classifier A would be chosen to discriminate between classes 1 and 2, as it is more accurate when compared to classifier B, i.e. its circle has a bigger area, denoting a greater number of correctly classified examples. Let us now discuss how appropriate the choice of classifier A (made by the greedy approach) is in each of the different scenarios shown in Figures 3 and 4, taking into account the interactions between classifiers A and C, and between B and C, in the context of the class tree shown in Figure 5.

Recall that in the TDDC approach an example is correctly assigned to class 1.1 or 1.2 if and only if the two following events occur: the example is correctly classified by the root classifier (A or B); and the example is correctly classified by classifier C. Therefore, the individual accuracy of each classifier is not necessarily the most important factor when selecting a candidate classifier; rather it is the number of examples correctly classified by *both* the parent and child classifiers (the intersection between their sets of correctly classified examples). In the case of Figure 5, in order to maximise the classification accuracy at the leaf class nodes 1.1 and 1.2, if $|A \cap C| > |B \cap C|$

then classifier A should be chosen; if it is not, B should be chosen. For this reason, the greedy approach produces a good selection in the case of Figure 4, where $|A \cap C| > |B \cap C|$. However, the greedy approach would not produce an optimal selection in the case of Figure 3. This is due to the fact that although A has a greater area (higher accuracy) in Figure 3, $|B \cap C| > |A \cap C|$.

4   Global Search-Based Classifier Selection with a Particle Swarm Optimisation/Ant Colony Optimisation Algorithm

Given the discussion in the previous section it is quite clear that there is a potential to improve the classification accuracy of the entire classifier tree by using a more "intelligent" classifier selector – a classifier selector that (unlike the greedy one) takes into account interaction among classifiers at different classifier nodes. As there is an obvious objective function to be optimised – the classification accuracy of the entire TDDC tree on the validation set – and also a collection of elements whose optimal combination has to be found – the type of classifier at each classifier node, it seems appropriate to use a combinatorial optimisation algorithm.

We propose to optimise the selection of a classifier at each classifier node with a PSO/ACO algorithm, adapted from the PSO/ACO algorithm described in [4] [5]. The choice of this algorithm was motivated by the following factors. Firstly PSO/ACO has been shown to be an effective classification-rule discovery algorithm [4] [5] across a wide variety of data sets involving mainly nominal attributes. Secondly, the PSO/ACO algorithm can be naturally adapted to be used as a classifier selector, where instead of finding a good combination of attribute-values for a rule, it finds good combinations of classifiers for all the nodes of the classifier tree. This is because a combination of classifiers is specified by a set of nominal values (types of classification algorithms). Due to size restrictions this section assumes the reader is familiar with standard PSO [6] and ACO algorithms [7].

A hybrid (PSO/ACO) method was developed to discover classification rules from categorical (nominal) data [4] [5]. In essence, this algorithm works with a population of particles. Each containing multiple pheromone vectors – each pheromone vector is used to probabilistically decide which value of a nominal attribute is best in each dimension of the problem's search space. In the original PSO/ACO for discovering classification rules these dimensions correspond to predictor attributes of the data being mined, so there is one pheromone vector for each nominal attribute. The entries in each individual pheromone vector correspond to possible values the attribute can take, and each pheromone value denotes the "desirability" of including the corresponding attribute value in a rule condition. We now describe in detail how this algorithm was adapted to act as a classifier selector, rather than discovering classification rules.

To optimise the classifier selection at each classifier node the problem must be reduced to a set of dimensions and possible values in each dimension. Hence, in the proposed PSO/ACO for classifier selection each decoded particle (candidate solution) consists of a vector with n components (dimensions), as follows:

$$Decoded\ Particle = w_1, w_2, \ldots, w_n$$

Where w_d ($d = 1, \ldots, n$) is the classifier selected at the dth classifier node in the TDDC tree and n is the number of classifier nodes in the tree. Each w_d can take one of the nominal (classifier ids) values $c_1,...c_k$ where k is the number of different candidate classifiers at each node.

It must also be possible to assess how good an individual solution created from an individual particle is. To do this the validation set is classified by the TDDC tree composed of the classifiers specified by the particle, and that tree's average classification accuracy (the mean of the accuracy from each class level) on the validation set is taken. The mean classification accuracy across all the class levels is used as the "fitness" (evaluation) function for evaluating each particle's quality.

Note that the only increase in computational time for this approach (over the greedy selective approach) is in the time spent classifying examples at each fitness evaluation. The classifiers are trained using the same data at each fitness evaluation and so can be cached and reused without the need for retraining.

```
Initialize population
REPEAT for MaxInterations
    FOR every particle P
        /* Classifier Selection */
        FOR every dimension w_d in P
            Use fitness proportional selection on pheromone vector
            corresponding to w_d to choose which state (classifier
            id) c_1,..c_k should be chosen for this w_d
        END FOR
        Construct a classifier tree by using the classifiers se-
        lected from the particle's pheromone vectors
        Calculate fitness F of this set of classifiers w_1,..w_n
        /* Set the past best position */
        IF F > P's best past combination's (P_b) fitness F_b
            F_b = F
            P_b = the current combination of classifiers w_1,..w_n
        END IF
    END FOR
    FOR every particle P
        Find P's best Neighbour Particle N according to each
        neighbour's best fitness (F_b)
        FOR every dimension w_d in P
            /* Pheromone updating procedure */
            f = N's best fitness F_b
            y = N's best state P_b in dimension d
            /* Add an amount of pheromone proportional to f to the
            pheromone entry for particle P corresponding to y (the
            best position held by P's best Neighbour) */
            τ_pdy = τ_pdy + (f × â)
            Normalize τ_pd
        END FOR
    END FOR
END REPEAT
```

Pseudocode 1. The Hybrid PSO/ACO Algorithm for Classifier Selection

Pseudocode 1 shows the hybrid PSO/ACO algorithm for classifier selection. At each iteration each pheromone vector for each particle produces a state in a probabilistic manner. That is, the probability of choosing a given classifier $(c_1,..c_k)$ for a given classifier node $(w_1,..w_n)$ is proportional to the amount of pheromone (a number between 0 and 1) in the corresponding entry in the corresponding pheromone vector

(τ_{pd} is the pheromone vector corresponding to particle P and classifier node d), see Figure 6. More precisely, the selection of a classifier at each classifier node is implemented by a fitness proportional (roulette-wheel) selection mechanism [8].

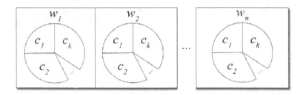

Fig. 6. An encoded particle with n dimensions, each with k classifier ids

Figure 6 shows an encoded particle P. Each section labelled $c_1, c_2, .. c_k$ (in each dimension $w_1, w_2, .., w_n$) represents an amount of pheromone. The probability of choosing each classifier c_i ($i=1,..,k$) in each dimension w_d ($d=1,..,n$) is proportional to the amount of pheromone (τ) in the corresponding pheromone entry τ_{pdi}.

The "decoded" state is then evaluated, and if it is better than the previous personal best state (P_b), it is set as the personal best state for the particle. A particle finds its best neighbour (N) according to the fitness of each neighbour's best state (P_b). In this paper the particles are arranged in a Von-Neumann topology [6], so that each particle has four neighbours.

A slightly different pheromone updating approach is taken with the PSO/ACO algorithm for classifier selection when compared to the PSO/ACO algorithm for rule discovery. As detailed in the pheromone updating procedure in Pseudocode 1, the approach simply consists of adding an amount of pheromone proportional to f to the pheromone entry corresponding to τ_{pdy}. Where f is the fitness of the best neighbour's best state, y is the best neighbour's best state ($c_1, .. c_k$) in the particular dimension d ($w_1, .. w_n$) and P is the current particle. Although not used in this paper the amount of pheromone added can be modified to slow down (or speed up) convergence, this is achieved using the constant $\acute{a}$. The closer this constant is set to 0 the slower the convergence achieved. The pheromone vectors are normalised after pheromone has been added, so that the pheromone entries of each pheromone vector add up to 1.

5   Experimental Setup

5.1   Bioinformatics Data Sets

The hierarchical classification methods discussed above were evaluated in four challenging datasets involving the prediction of protein function. The protein functional classes to be predicted in these data sets are the functional classes of GPCRs (G-Protein-Coupled Receptors). GPCRs [9] are proteins involved in signalling. They span cell walls so that they influence the chemistry inside the cell by sensing the chemistry outside the cell. More specifically, when a ligand (a substance that binds to a protein) is received by the part of the GPCR on the outside of the cell, it (usually) causes an attached G-protein to activate and detach. GPCRs are very important for medical applications because 40%-50% of current drugs target GPCR activity [9].

Predicting GPCR function is particularly difficult because the types of function GPCRs facilitate are extremely varied, from detecting light to managing brain chemistry.

The GPCR functional classes are given unique hierarchical indexes by [10]. The GPCRs, examples (proteins) have up to 5 class levels, but only 4 levels are used in the datasets created in this work, as the data in the 5th level is too sparse for training – i.e., in general there are too few examples of each class at the 5th level. In any case, it should be noted that predicting all the first four levels of GPCR's classes is already a challenging task. Indeed, most works on the classification of GPCRs limit the predictions to just one or two of the topmost class [11], [12], [13], [14].

The data sets used in our experiments were constructed from data in UniProt [15] and GPCRDB [10]. UniProt is a well known biological database, containing sequence data and a rich annotation about a large number of proteins. It also has cross-references for other major biological databases. It was extensively used in this work as a source of data for creating our data sets. Only the UniProtKB/Swiss-Prot was used as a data source, as it contains a higher quality, manually annotated set of proteins. Unlike Uniprot, GPCRDB is a database specialised on GPCR proteins.

We performed experiments with four different kinds of predictor attributes, each of them representing a kind of "protein signature", or "motif", namely: FingerPrints from the Prints [17] database, Prosite patterns [16], Pfam [18] and Interpro entries [19]. The four GPCR data sets each use predictor attributes from one of either the Prints, Prosite, Interpro or Pfam databases. They also contain two additional attributes, namely the protein's molecular weight and sequence length.

Any duplicate examples (proteins) in a data set are removed in a pre-processing step, i.e., before the hierarchical classification algorithm is run, to avoid redundancy. If there are fewer than 10 examples in any given class in the class tree that class is merged with its parent class. If the parent class is the root node, the entire small class is removed from the data set. This process ensures there is enough training and test data per class to carry out the experiments. (If a class had less than 10 examples, during the 10-fold cross-validation procedure there would be at least one iteration where there would be no example of that class in the test set).

After data pre-processing, the final datasets used in the experiments have the numbers of attributes, examples (proteins) and classes per level (expressed as level 1/ level 2/level 3/level 4) indicated in Table 1.

Table 1. Main characteristics of the datasets used in the experiments

	GPCR/Prints	GPCR/Prosite	GPCR/Interpro	GPCR/Pfam
#Attributes	283	129	450	77
#Examples	5422	6261	7461	7077
#Classes	8/46/76/49	9/50/79/49	12/54/82/50	12/52/79/49

5.2 Data Set Partitioning and Algorithmic Details

The data sets were split into two main subsets at each iteration of the 10-fold cross validation process, one test set and one training set. The test set is used to assess the performance of the approach in question; therefore the true class of each test example remains unseen during the training process, only to be revealed to measure the predictive accuracy of the approach. The training set is split into a further two subsets. Firstly 75% of the training set was used as the building set; this building set is used to train the classifiers. Secondly the validation set, which consists of the remaining 25% of the training examples. The validation set is used to compute the quality of the classifiers, and so particle fitness in the PSO/ACO algorithm. After the best solution (according to accuracy in the validation set) has been found in a single PSO/ACO run, the classifiers at every classifier node specified in that best particle are trained using the entire training set. This procedure attempts to maximise the individual classifier's accuracy and so the final accuracy in the test set.

As a baseline it is important to evaluate the proposed method by comparing its predictive accuracy with the predictive accuracy of the greedy selective top-down approach. The baseline should also include each of the individual classification algorithms used in the greedy selective top-down approach. Therefore the first experiments are to build standard TDDC trees using one type of classification algorithm throughout.

The classification algorithms used in the experiments presented in this paper were implementations from the WEKA [20] package. These algorithms were chosen to include a diverse set of paradigms, while having high computational efficiency:

- HyperPipes is a very simple algorithm that constructs a "hyperpipe" for every class in the data set; each hyperpipe contains each attribute-value found in the examples from the class it was built to cover. An example is classified by finding which hyperpipe covers it the best.
- NaiveBayes uses Bayes' theorem to predict which class an example most likely belongs to, it is naïve because it assumes attribute independence.
- J48 is a decision tree algorithm, being WEKA's modified version of the very well known C4.5 algorithm.
- ConjunctiveRule is another very simple algorithm that only produces two rules to classify the entire data set. A "default" rule is produced that predicts the class with the greatest numbers of records in the training set. The other rule is constructed using information gain to select attribute-values for the antecedent.
- BayesNet uses a Bayesian network to classify examples and can theoretically completely take into account attribute dependency.

Although some of these algorithms are clearly more advanced than the others, all were selected for some classifier nodes by the classifier selection method (greedy approach or PSO/ACO) during training, confirming that all of them perform best in certain circumstances. All experiments were performed using 10-fold cross validation [20] with $á$ set to 1 for the PSO/ACO algorithm.

6   Computational Results

The predictive accuracy for each method (the five baseline clasifiers used thoughout the TDDC tree, the greedy and PSO/AOCO methods for classifier selection) are shown in Tables 2 through 5 for each dataset. The values after the "±" symbol are standard deviations (calculated using the WEKA statistics classes). Tables 2 through 5 are shown for the sake of completeness, but, to simplify the analysis (and due to paper size restrictions) we focus mainly on a summary of the results (Table 6). Table 6 shows the summary of the number of cases where there is a statistically significant difference in the predictive accuracy of the 2 methods according to the WEKA corrected two-tailed student t-test (with a significance level 1%). Each cell shows the number of times the labelled approach (Greedy or PSO/ACO) significantly beats the baseline classification algorithm (HP – HyperPipes, NB – NaiveBayes, CR – ConjunctiveRule, BN – Bayes-Net), in each data across all four class levels. Totals across all data sets are shown at the bottom of the table.

Table 2. Percentage accuracy for each approach in the Prints data set

TDDC Type	Percentage accuracy at each level in the class hierarchy			
	1st	2nd	3rd	4th
HyperPipes	90.76±0.34	76.79±0.55	49.99±1.1	75.42±2.11
NaiveBayes	87.74±0.71	72.72±1.11	41.3±0.99	63.85±1.89
J48	91.68±0.51	83.35±1.0	58.34±1.26	85.14±1.8
ConjunctiveRule	80.16±0.31	49.63±0.46	17.03±0.84	24.8±0.87
BayesNet	88.34±1.39	77.41±1.25	48.0±0.93	74.53±2.94
Greedy	91.68±0.51	83.06±0.88	58.21±1.23	84.66±2.09
PSO/ACO	91.59±0.52	82.67±1.13	57.99±1.52	84.8±2.34

Table 3. Percentage accuracy for each approach in the Interpro data set

TDDC Type	Percentage accuracy at each level in the class hierarchy			
	1st	2nd	3rd	4th
HyperPipes	83.74±1.14	73.77±1.01	48.21±0.95	82.62±2.5
NaiveBayes	87.88±0.59	74.78±0.78	38.59±1.07	51.25±1.85
J48	90.36±0.34	80.68±0.66	51.06±0.93	79.86±2.68
ConjunctiveRule	73.68±0.18	47.73±0.48	17.76±0.47	24.84±0.68
BayesNet	89.18±0.67	78.99±0.83	46.4±0.94	67.3±2.62
Greedy	90.36±0.34	80.41±0.81	54.36±1.33	83.58±2.46
PSO/ACO	90.36±0.34	80.4±0.78	54.43±1.27	84.24±2.27

Table 4. Percentage accuracy for each approach in the Pfam data set

TDDC Type	Percentage accuracy at each level in the class hierarchy			
	1st	2nd	3rd	4th
HyperPipes	92.02±0.44	25.4±0.75	9.8±0.82	4.58±1.22
NaiveBayes	89.59±0.72	59.23±1.41	19.6±1.43	16.27±2.39
J48	92.98±0.48	70.77±1.39	37.03±1.07	48.97±3.98
ConjunctiveRule	75.55±0.13	51.4±0.53	13.49±2.0	6.97±4.63
BayesNet	90.35±1.1	62.7±1.45	23.25±1.46	23.43±2.42
Greedy	92.98±0.48	70.54±1.29	36.97±1.2	48.24±3.55
PSO/ACO	92.98±0.48	70.5±1.35	36.97±1.21	48.5±3.58

Table 5. Percentage accuracy for each approach in the Prosite data set

TDDC Type	Percentage accuracy at each level in the class hierarchy			
	1st	2nd	3rd	4th
HyperPipes	82.14±0.71	46.03±1.28	23.1±1.62	32.16±2.82
NaiveBayes	85.34±1.14	60.63±1.25	24.86±1.3	23.94±2.11
J48	84.71±0.57	61.02±1.12	29.31±1.63	39.58±3.35
ConjunctiveRule	78.68±0.15	41.38±0.25	14.79±0.45	10.0±0.89
BayesNet	85.93±0.88	62.17±1.06	26.68±1.35	31.14±2.47
Greedy	85.93±0.88	62.54±0.91	31.46±1.25	40.73±4.21
PSO/ACO	85.93±0.88	62.8±1.33	32.18±1.48	43.11±3.71

Table 6. Summation of the number of statistically significant resutlts

Dataset	Classif. Selection Approach	Classification Algorithm				
		HP	NB	J48	CR	BN
GPCR/Prints	Greedy Selective	4	4	0	4	4
	PSO/ACO	4	4	0	4	4
GPCR/InterPro	Greedy Selective	3	4	1	4	4
	PSO/ACO	3	4	2	4	4
GPCR/Pfam	Greedy Selective	4	4	0	4	4
	PSO/ACO	4	4	0	4	4
GPCR/Prosite	Greedy Selective	4	2	1	4	2
	PSO/ACO	4	3	3	4	2
Totals	Greedy Selective	15	14	2	16	14
	PSO/ACO	15	15	5	16	14

Both the greedy and PSO/ACO approach for classifier selection were very success-ful in improving predictive accuracy with respect to four of the base classification algorithms (HP, NB, CR, BN), as shown by the totals in Table 6. These two ap-proaches were less successful in improving accuracy with respect to J48, but even in this case the classifier selection approaches improved upon J48's accuracy several times, whilst never decreasing upon J48's accuracy.

The PSO/ACO classifier selection approach significantly improves upon the per-formance of the greedy approach in four cases overall. PSO/ACO improves on the performance of J48 in five cases, three more than the greedy approach. These im-provements are in the third and fourth level of the Prosite dataset and there is also an improvement in the InterPro dataset at the fourth level. As J48 is the hardest classifi-cation algorithm to beat, these results show the most difference. However, the PSO/ACO algorithm also scores better against NaiveBayes when compared to the greedy approach in one case – in the Prosite dataset at the second class level.

The results imply that both the PSO/ACO algorithm and greedy approaches benefit more from more "difficult" data sets. The data set in which the base classification algorithms perform worst is the Prosite data set. This data set also yields the biggest improvement in accuracies when using the greedy (1 significant win over J48), and more so the PSO/ACO (3 significant wins over J48) approach. Indeed for either of these approaches to increase predictive accuracy above that of a base classifier, the base classifier must make an error that is not made by another base classifier. The more mistakes made by a certain classification algorithm (due to a more difficult data set) the higher the probability of another classification algorithm not making the same set of mistakes. Furthermore, it was observed that overfitting is sometimes a limiting factor with the PSO/ACO approach, since increases in validation set accuracy (over the baseline classification algorithms) did not always result in a similar increase in test set accuracy.

7 Conclusions and Future Research

Our experiments show that both the greedy and PSO/ACO approaches for classifier selection significantly improve predictive accuracy over the use of a single fixed algorithm throughout the classifier tree, in the majority of cases involving our data sets. Overall, the PSO/ACO approach was somewhat more successful (significantly better in four cases) than the greedy approach. We believe that the use of a more advanced approach (as discussed in this paper) is more appropriate in more difficult data sets, where classification algorithms are more likely to make mistakes. Estimating a priori how likely a classification algorithm is to make a mistake is an open problem and this topic is left for future research. In this work the proposed PSO/ACO was compared only with Secker et al's greedy selective approach, so one direction for future research is to compare the PSO/ACO with another population-based meta-heuristics for optimisation, e.g. evolutionary algorithms.

References

1. TrEMBL. Visited (June 2007), http://www.ebi.ac.uk/swissprot/sptr_stats/full/index.html
2. Clare, A., King, R.D.: Machine learning of functional class from phenotype data. Bioinformatics 18(1), 160–166 (2007)
3. Secker, A., Davies, M.N., Freitas, A.A., Timmis, J., Mendao, M., Flower, D.: An Experimental Comparison of Classification Algorithms for the Hierarchical Prediction of Protein Function. Expert Update (British Computer Society – Specialist Group on Artificial Intelligence Magazine) 9(3), 17–22 (2007)
4. Holden, N., Freitas, A.A.: Hierarchical Classification of G-Protein-Coupled Receptors with a PSO/ACO Algorithm. In: Proc. IEEE Swarm Intelligence Symposium (SIS 2006), pp. 77–84. IEEE, Los Alamitos (2006)
5. Holden, N., Freitas, A.A.: A hybrid PSO/ACO algorithm for classification. In: Proc. of the GECCO-2007 Workshop on Particle Swarms: The Second Decade, pp. 2745–2750. ACM Press, New York (2007)
6. Kennedy, J., Eberhart, R.C., Shi, Y.: Swarm Intelligence. Morgan Kaufmann/ Academic Press (2001)
7. Dorigo, M., Stützle, T.: Ant Colony Optimization. MIT Press, Cambridge (2004)
8. Eiben, A.E., Smith, J.E.: Introduction to Evolutionary Computing. Natural Computing Series, 2nd edn. (2007)

9. Fillmore, D.: It's a GPCR world. Modern drug discovery 11(7), 24–28 (2004)
10. GPCRDB (2007), http://www.gpcr.org/
11. Bhasin, M., Raghava, G.P.: GPCRpred: An SVM-based method for prediction of families and subfamilies of G-protein coupled receptors. Nucleic Acids Res. 1(32 Web Server issue), 383–389 (2004)
12. Guo, Y.Z., Li, M.L., Wang, K.L., Wen, Z.N., Lu, M.C., Liu, L.X., Jiang, L.: Classifying G protein-coupled receptors and nuclear receptors on the basis of protein power spectrum from fast Fourier transform. Amino Acids 30(4), 397–402 (Epub, 2006)
13. Karchin, R., Karplus, K., Haussler, D.: Classifying G-protein coupled receptors with support vector machines. Bioinformatics 18(1), 147–159 (2002)
14. Papasaikas, P.K., Bagos, P.G., Litou, Z.I., Hamodrakas, S.J.: A novel method for GPCR recognition and family classification from sequence alone using signatures derived from profile hidden Markov models. SAR QSAR Environ Res 14(5-6), 413–420 (2003)
15. UniProt (June 2007), http://www.expasy.UniProt.org/
16. Hulo, N., Bairoch, A., Bulliard, V., Cerutti, L., De Castro, E., Langendijk-Genevaux, P.S., Pagni, M., Sigrist, C.J.A.: The PROSITE database. Nucleic Acids Res. 34, D227–D230 (2006)
17. Attwood, T.K.: The PRINTS database: A resource for identification of protein families. Brief Bioinform., 252–263 (2002)
18. Bateman, A., Coin, L., Durbin, R., Finn, R.D., Hollich, V., Griffiths-Jones, S., Khanna, A., Marshall, M., Moxon, S., Sonnhammer, E.L.L., Studholme, D.J., Yeats, C., Eddy, S.R.: The Pfam protein families database. Nucleic Acids Research 32(Database-Issue), 138–141 (2004)
19. Mulder, N.J., et al.: New developments in the InterPro database. Nucleic Acids Res. 35(Database Issue), D224–D228 (2007)
20. Witten, I.H., Frank, E.: Data Mining: Practical Machine Learning Tools and Techniques, 2nd edn. Morgan Kaufmann, San Francisco (2005)

Protein Interaction Inference Using Particle Swarm Optimization Algorithm

Mudassar Iqbal, Alex A. Freitas, and Colin G. Johnson

Centre for Biomedical Informatics and Computing Lab., University of Kent,
Canterbury, UK
{mi26,a.a.freitas,c.g.johnson}@kent.ac.uk

Abstract. Many processes in the cell involve interaction among the
proteins and determination of the networks of such interactions is of im-
mense importance towards the complete understanding of cellular functions. As the experimental techniques for this purpose are expensive
and potentially erroneous, there are many computational methods being
put forward for prediction of protein-protein interactions. These methods use different genomic features for indirect inference of protein- protein interactions. As the interaction among two proteins is facilitated
by domains, there are many methods being put forward for inference
of such interactions using the specificity of assignment of domains to
proteins. We present here an heuristic optimization method, particle
swarm optimization, which predicts protein-protein interaction by using
the domain assignments information. Results are compared with another
known method which predicts domain interactions by casting the problem of interactions inference as a maximum satisfiability (MAX-SAT)
problem.

1 Introduction

Computational inference of protein-protein interactions is an interesting and
challenging area of research in modern biology. Computational methods infer
potential interactions using one or more genomic features related to the protein
pairs as predictor attributes. Many genomic experiments have produced some
high quality information regarding genes/proteins which is not directly related
to their interaction but could potentially be used for such a purpose.

Many computational methods use a single type of genomic data to predict protein interactions,e.g, using similarity in phylogenetic profiles, gene fusion methods, or the hypothesis involving co-expression or co-localization of interacting
partners. Other methods integrate different genomic features using a variety of
machine learning methods to infer new protein-protein interactions. In [1,2,3],
one can find a few recent reviews regarding experimental and computational
methods for protein-protein interaction prediction.

An important area under focus in many research projects is to infer protein
interactions by looking at their domain compositions. Domains are evolution-
arily conserved sequence units which are believed to be the responsible for the

E. Marchiori and J.H. Moore (Eds.): EvoBIO 2008, LNCS 4973, pp. 61–70, 2008.

interactions among the proteins to which they belong. There are many different methods which infer protein interactions using information on their domain composition. A protein pair is thought to be physically interacting if at least one of their constituent domain pair interacts. Most of the proteins in organisms like *S. Cerevisiae* are assigned one or more domains and information about the domains pairs in high confidence experimentally determined protein interaction data sets can be used to infer domain-domain and hence, protein-protein interaction. As there are no specific domain interaction data available, many methods have been developed for finding potential domain interaction from available experimentally determined high confidence protein-protein interaction datasets and then that information is used to predict back the novel protein-protein interactions as well [4,5,6,7]. In other words, these methods infer domain-domain interactions from protein protein interactions and use these inferred domain interactions to predict new protein-protein interactions, given the composition of domains in those proteins.

In a recent work [8,9], a combinatorial approach is proposed for the inference of protein interactions using domain information. In the framework they use, this inference problem is presented as a satisfiability (more precisely MAX-SAT) problem, as explained in detail in Section 2, which is then solved using linear programming method by relaxing some of constraints of the original MAX-SAT problem.

In this work we propose the use of particle swarm optimization to solve this maximum satisfiability problem, using the problem formulation based on the one originally proposed in [8] and we also implement the technique employed by them to compare the results. Particle swarm optimization (PSO) is a population based heuristic optimization technique [10,11], inspired by the social behavior of bird flocking or fish schooling [13]. It has been successfully used for optimizing high dimensional complex functions, mostly in continuous application domains. A good recent review about the different developments and applications on PSO can be found in [14].

This paper is organized as following. Section 2 details the formulation of the protein interaction problem into a MAX-SAT problem, as it is done originally in [8]. Section 3 proposes the use of a Particle Swarm Optimization algorithm (PSO) for this problem and discusses the related design issues for the use of PSO. In section 4, data sets about the domain assignments and protein interactions used in the experiments are described, and computational results are reported. Finally, Section 5 concludes the paper.

2   Protein Interaction Inference as MAXSAT Problem

We follow the problem formulation as is done in [8,9], based on the hypothesis that a protein pair is interacting if and only if at least one pair of their domains (one from each protein) interact and non-interacting otherwise. We denote $P = \{p_1, p_2, ..., p_M\}$ as a set of proteins, $D = \{d_1, d_2, ..., d_N\}$ as a set of domains and Ω_{ij} as the set of unique domain pairs contained in a protein pair (p_i, p_j). Let us consider two variables defining protein-protein and domain-domain interactions

$$P_{ij} = \begin{cases} 1 \text{ if proteins } p_i \text{ and } p_j \text{ interact} \\ 0 \qquad\qquad \text{Otherwise} \end{cases}$$

$$D_{nm} = \begin{cases} 1 \text{ if domains } d_n \text{ and } d_m \text{ interact} \\ 0 \qquad\qquad \text{Otherwise} \end{cases}$$

Given the domain-domain interactions, we can predict the protein pairs interacting or non-interacting depending upon their corresponding domain pairs as:

$$P'_{ij} = \bigvee_{d_{nm} \in \Omega_{ij}} D_{nm} \tag{1}$$

Where the true outcome of this logical operation means the corresponding protein pair is interacting (i.e, 1) and false means non-interacting (i.e., 0). Using this relationship one needs to find the best assignment of 1's and 0's to the domain variables which best represents the data, i.e., a SAT (satisfiability) assignment satisfying all interacting and non-interacting protein pairs. As we know there are many false positives and false negatives in experimental data, such an assignment is not possible to find, so we will look for an assignment which satisfies the maximum number of relationships (clauses), which is known as the MAX-SAT problem. These problems are very difficult to solve in general and their exact solutions are not possible in general. This problem is solved in [8] using linear programming by relaxing some of the constraints as described in equations 2 and 3. The following linear program was formulated by relaxing the binary constraints on the variables.

$$\text{Minimize} \qquad \sum_{ij} |P_{ij} - P'_{ij}|$$

$$\text{Subject To:} \qquad \sum_{d_{nm} \in \Omega_{ij}} D_{nm} \geq P_{ij} \quad \forall (i,j) \tag{2}$$

$$0 \leq P'_{ij} \leq 1 \quad \forall (i,j)$$

$$0 \leq D_{nm} \leq 1 \quad \forall (n,m)$$

P_{ij} is 1 or 0 if the two proteins p_i and p_j interact or not respectively, according to experimental data. Equation 2 can also be expressed in the following form.

$$\text{Minimize} \qquad \sum_{P_{ij}=0} P'_{ij} - \sum_{P_{ij}=1} P'_{ij}$$

$$\text{Subject To:} \qquad \sum_{d_{nm} \in \Omega_{ij}} D_{nm} \geq P_{ij} \quad \forall (i,j) \tag{3}$$

$$0 \leq P'_{ij} \leq 1 \quad \forall (i,j)$$

$$0 \leq D_{nm} \leq 1 \quad \forall (n,m)$$

The real values obtained for variables P'_{ij} and D_{nm} after optimization represent the probabilities of they taking the integer value 1, and a threshold can be used to convert them back to binary.

3   Binary Particle Swarm Optimization Algorithm for Inference of Protein Interactions

Particle swarm optimization (PSO) is a population-based stochastic optimization technique developed by Eberhart and Kennedy in 1995 [10,11,12], inspired by the social behaviour of bird flocking or fish schooling.

PSO shares many similarities with evolutionary computation techniques such as Genetic Algorithms (GA). The system is initialized with a population of random solutions (particles) and searches for optima of the given objective function by iteratively updating the positions of those particles. However, unlike GA, PSO has no genetic operators such as crossover and mutation. In PSO, the potential solutions, called particles, fly through the problem space as they are attracted by the other particle positions in the neighbourhood representing good quality candidate solutions. An individual's neighbourhood may be defined in several ways,configuring somehow the "social network" of the individual. Several neighbourhood topologies exist (full, ring, star, etc.) depending on whether an individual interacts with all, some, or only one of the rest of the population.

PSO has shown promising results on many applications, especially continuous function optimisation. A good recent review of relevant research in this area can found in [14]. The basic idea of the proposed work here is to extend the application of PSO to a more challenging real world problem, namely the inference of protein interactions, which can be framed as an optimization problem (as discussed in section 2) given the assignment of domains to the proteins, where the goal is to find the network of interactions that best explains the given experimental dataset.

In the Binary version of PSO individual components of a candidate solution (particle) are not real valued, rather 1 or 0, and velocity is interpreted as proportional likelihood, which is used in the logistic function to generate a particle's binary positions, i.e.

$$v_{id}^{t+1} = w * v_{id}^{t} + c_1 * \phi_1 * (p_{id}^{t} - x_{id}^{t}) + c_2 * \phi_2 * (p_{gd}^{t} - x_{id}^{t}) \tag{4}$$

$$x_{id}^{t+1} = 1 \quad \text{if} \quad \phi_3 < \frac{1}{1 + e^{-k*v_{id}^{t+1}}} \quad \text{else} \quad 0 \tag{5}$$

Where $x_{id} \in \{0, 1\}$ is the value for the dth dimention of particle i and v_{id} is the velocity, which is clamped between to a maximum value, $|V_{max}|$. p_{id} and p_{gd} are the best positions in the dth dimenstion of particle i and its neighbourhood's best particle g respectively. t is the iteration index, and w is the inertia weight, determining how much of the previous velocity of the particle is preserved. This plays the role of balancing the global and local search ability of PSO [15]. Parameter k in the logistic function is a positive constant which controls the shape of the curve. c_1, c_2 are two positive acceleration constants while ϕ_1, ϕ_2 and ϕ_3 are three uniform random numbers sampled from $U(0, 1)$. For the velocity update equation, in terms of social psychology as a metaphor, the second part of the right hand side of the velocity update equation represents the private thinking by itself; the third part is the social part, which represents the cooperation among the individuals.

3.1   Solution Representation and Objective function

Each particle represents a candidate solution to the inference problem. The position vector of particle m is $X_m = \{d_{ij}\}$ where index ij runs over all unique domains pairs in the data, i.e, a particle consists of a binary string where each bit refers to a distinct unique domain pair in the training data. These are the bits which the particle will try to optimise during the course of evolution by updating its velocity and position according to equations 4 and 5, by interacting with its neighbourhood. The *gbest* version of binary PSO is used for these experiments. Each protein pair expressed in terms of its constituent domain pairs is a clause from the point of view of logic. The objective is to maximize the number of satisfied clauses or equivalently minimize the number of unsatisfied clauses. Let us define a variable P'_{ij} for each protein pair (p_i, p_j) to indicate whether it is predicted interacting or not according to the given assignment of domain pairs by some particle (solution). The objective function can be expressed as a minimization problem.

$$\text{Min} \quad f = \sum_{ij} \mid P_{ij} - P'_{ij} \mid$$

$$\text{Such that} \quad P'_{ij} = \bigvee_{d_{nm} \in \Omega_{ij}} D_{nm} \qquad (6)$$

4   Experimental Design and Results

4.1   Protein-Protein Interaction and Domain Assignment Data

We obtained domain assignments from SUPERFAMILY data base [16,17]. Superfamily database is a library of Hidden Markov Models that represents all proteins of known structure. These models are used to annotate the sequence of over 50 genomes. For *S. Cerevisiae* organism there exists 3346 sequences with at least one domain assignment, which is about 50% of the total sequences. In total 4681 domains are assigned, and there are 685 superfamily domains with at least one assignment.

We obtained the *sS. Cerevisiae* interaction data set from DIP (Data base of Interacting Proteins [18]). We obtained nearly 5000 high confidence positive interaction in DIP which is a subset of experimentally determined interaction in DIP, called CORE. Negative interactions are hard to find. As used by many researchers in this field (e.g. [20],[21]), we use protein pairs being defined as non-interacting if they are not in same cellular compartment. This gives us many hundreds of thousand of protein pairs which are not co-localized. This is a huge data set compared with the number of positives, so we randomly sample some negatives from this pool of possible negatives, in order to obtain a more balanced class distribution for the classification algorithm. Then we only want to keep those positive or negative pairs which have at least one domain assignment for each protein in the pair in the superfamily database, as there are some proteins which do not have any significant domain assignment. This process reduces our data set of positive interactions to 3070 pairs. We also created two sets of negative examples, one with the same number of negative examples (protein pairs) as the number of positive examples, i.e., 3070, while the other with 4000 negatives. In our experiments, we will call the first dataset containing 3070 positive interactions and 3070 negative interactions as *data1*, while the other data set contraining 3070 positive interactions and 4000 negative interactions will be called as *data2*.

4.2   PSO Parameters

We rely on the standard PSO parameters settings [13]. The two constants c_1 and c_2 are set to 2.0, while parameter k in the logistic function is set to 5.0. Maximum velocity (V_{max}) is set to 4.0 and individual particle's velocities in each dimension are initialized uniformly between $-V_{max}$ and $+V_{max}$. An important issue regarding the initialization of swarm is analyzed in detail.

PSO Initialization: A Data-Driven Approach. We have to decide a starting configuration for PSO, e.g., the probability of a particle taking the value 1 (or 0) in each dimension, for all particles in the swarm. Usually the population in PSO is initialized completely randomly, but PSO has dependance on initial conditions (in this case, how many 1's or 0's we put into the system at the start). Hence, one needs to find an objective and consistent way to decide the initial configuration,i.e., initial number of 1's (or zeros for that matter) in the system.

In our case, one solution to this issue is to use the domain assignment information apriori to calculate the initialization probability (of being 1 or 0) for each domain separately, and use that to probabilistically assign 1 or 0 value to all the domain pair variables, that is, for every domain pair ij, we calculate the counts of being in interacting protein pairs and non-interacting protein pairs, denoted C_p^{ij} and C_n^{ij} respectively. We have the probability of being in state 1 given by Eq. 7.

$$F_{ij} = \frac{C_p^{ij}}{(C_p^{ij} + C_n^{ij})} \qquad (7)$$

Now for each domain pair ij, we generate a random number r from a uniform probability distribution $U(0,1)$. If this number is less than F_{ij}, we assign 1 to that domain else 0. We use this scheme for all the experiments done using PSO.

4.3   Cross-Validation: Predicting Domain-Domain Interactions

In order to solve the linear program formulated in equation 4, as originally done in [8], we used GNU Linear Programming Kit [19](version 4.7). We used the interior point method which is a polynomial time linear programming algorithm within GNU Linear Programming Kit. The P'_{ij} values for protein pair i, j are calculated by summing over all domain pair variables $D_{nm} \in \Omega_{ij}$ and dividing by the number of domain pairs each protein pair contains in order to keep it within the bounds set in the linear program in equation 3. Since the variables

D_{nm} are not binary now, we used a threshold of 0.6 to convert them back to binary form, in accordance with the original work.

An important observation about our data sets is that many of the domain pairs occur either in positive protein pairs or negatives protein pairs only. This probably has something to do with our composition of the negative data. So, in the case of PSO, we in fact exclude those domain pairs from the PSO update process, i.e., they are fixed as either zero or one, depending upon which class of protein pairs they occur, but indeed they are included while evaluating the objective function in equation 6. This does not affect the prediction accuracy, but it greatly improves the running time of the algorithm, since the algorithm has fewer unknown variables to optimize.

For both data sets, we do a 10-fold cross validation procedure. For each experiment, we divide the data (for both positive and negative classes separately) randomly in ten equal folds. Each time we use nine out of ten folds as training and the remaining one fold as a test. This process is repeated ten times each time using a different fold as the test set. For *data1* in the Tables 1 and 2, we used 100 particles and PSO was allowed to run for 500 iterations, while in the case of *data2*, the number of iterations was increased to 1000.

For each of the 10 iterations of cross-validation procedure, we infer the domain pair interactions from the training set and use those interactions to predict protein pair interactions in the test set by using the relationship in equation 1, which can also be expressed in the following algebraic form.

$$P'_{ij} = 1 - \prod_{d_{nm} \in \Omega_{ij}} (1 - D_{nm}) \tag{8}$$

Tables 1 and 2 report the average results over all 10 cross-validation folds with corresponding standard deviations, for both datasets corresponding to the particle swarm optimization algorithm as well as the linear programming method (referred as LP in tables) respectively. *TPR* in the tables is defined as true positives over total number of positives and *FPR* is defined as false positives over total number of negatives in the data. Sensitivity is the same as *TPR* while

Table 1. Results for prediction of protein-protein interactions on test data, *data1*

Method	No. of 1's	Accuracy	TPR	FPR	Sensitivity*Specifity
PSO	1875 ± 11.06	0.826 ± 0.02	0.889 ± 0.015	0.289 ± 0.038	0.63±0.039
LP	1985 ± 9.68	0.81 ± 0.016	0.95 ± 0.01	0.45 ± 0.04	0.52±0.039

Table 2. Results for prediction of protein-protein interactions on test data, *data2*

Method	No. of 1's	Accuracy	TPR	FPR	Sensitivity*Specifity
PSO	1823 ± 13.32	0.81 ± 0.012	0.859 ± 0.017	0.253 ± 0.017	0.64±0.18
LP	1956 ± 13.42	0.78 ± 0.013	0.938 ± 0.014	0.437 ± 0.015	0.53±0.02

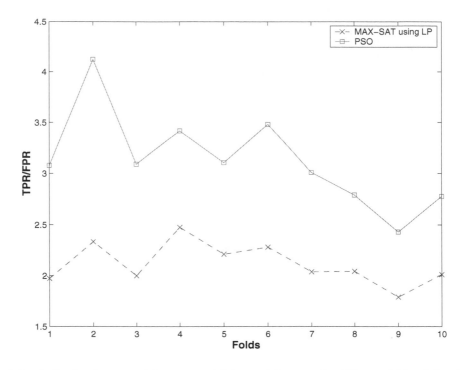

Fig. 1. Ratio of true positive rate and false positive rate for different folds of *data1*

Specificity is defined as $1 - FPR$. The performance of both methods is reported in terms of accuracy of prediction on test data, their corresponding true and false positive rates as well as the number of domain pairs predicted interacting (column "No. of 1's" in the tables). Protein pairs in test data which do not contain any domain pair from the training data were removed.

We can see from the Tables 1 and 2 that PSO produced better and more balanced results with a much lower rate of false positives. Results with two data sets, i.e., when we increase the proportion of negative examples from *data1* to *data2*, are not much different in the case of PSO, while they are significantly different in the case of the linear programming method. A statistical significance test (more precisely, a two-tailed student's *t*-test) was performed using the accuracy of both methods, and we obtained P-values for the paired t-test as 0.00073 and 0.0000026 at 95% confidence level corresponding to *data1* and *data2* respectively. The most probable explanation for these differences lies in the definition of the linear program in equation 3, which relaxes the constraint which eventually favours the positive interactions, hence much more false positive predictions. Fig. 1 shows the comparison between the two methods according to the true positive rate over false positive rate (TPR/FPR) for different folds (for data1). A qualitatively similar situation occurs for *data2*, and those results are not shown here for the sake of simplicity.

5   Conclusions

In this work, we have addressed an important bioinformatics problem, namely, the prediction of protein-protein interactions using information on their domain assignments. Particle swarm optimization is a relatively recent but very sucessful method in different optimization problems, but so far it has never been evaluated in the type of bionformatics problem addressed here. The problem has been cast as a combinatorial optimization problem, which allowed us to propose a novel use for a binary PSO algorithm. We have compared results with a known method which solve the same problem using linear programming techniques. Comparative results in terms of predictive accuracy on test data (unseen during training) show that PSO is a competitive optimizer in an application domain involving binary variables as well. We show that PSO not just achieves significantly better predictive accuracy overall but also reduces the false positive predictions.

As far as the prediction of protein-protein interaction in general is concerned, domain information might not be enough to determine completely the protein interactions, due to other possible factors. As a future research direction, it will be worth integrating this information with other features like RNA co-expression, etc., and to use data mining techniques for finding some associations between them which can be helpful in further understanding the mechanisms of protein and domain interactions.

Acknowledgements

Mudassar Iqbal acknowledges the support from the Computing Laboratory, University of Kent and the EPSRC grant GR/T11265/01 (eXtended Particle Swarms).

References

1. Benjamin, A., et al.: Deciphering Protein-Protein Interactions. Part I. Experimental Techniques and Databases. PLoS Comput Biol 3(3) e42 (2007)
2. Benjamin, A., et al.: Deciphering Protein-Protein Interactions. Part II. Computational Methods to Predict Protein and Domain Interaction Partners. PLoS Comput Biol 3(4) e43 (2007)
3. Valencia, A., Pazos, F.: Computational methods for the prediction of protein interactions. Current Opinion in Structural Biology 12, 368–373 (2002)
4. Riley, R., et al.: Inferring Protein Domain Interactions From Databases of Interacting Proteins. Genome Biology 6(R89) (2005)
5. Deng, M., et al.: Inferring Domain-Domain Interactions From Protin-Protein Interactions. Genome Res 12, 1540–1548 (2002)
6. Lee, H., et al.: An Integrated Approach to the Prediction of Domain-Domain Interactions. BMC Bioinformatics 7(269) (2006)
7. Li, X., et al.: Improving domain-based protein interaction prediction using biologically-significant negative dataset. International Journal of Data Mining and Bioinformatics 1(2), 138–149 (2006)

8. Zhang, Y., et al.: Protein Interaction Inference as a MAX-SAT Problem. In: Proceedings of IEEE Computer Society Conference on Computer Vision and Pattern Recognition, San Diego (2005)
9. Zhang, Y., et al.: Towards Inferring Protein Interactions: Challenges and Solutions. EURASIP Journal on Applied Signal Processing Article ID 37349 (2006)
10. Kennedy, J., Eberhart, R.: Particle swarm optimization. In: Proc. IEEE Int. Conf. on Neural Networks, pp. 1942–1948 (1995)
11. Eberhart, R., Kennedy, J.: A new optimizer using particle swarm theory. In: Proc. 6 th Int. Symposium on Micro Machine and Human Science, pp. 39–43 (1995)
12. Shi, Y., Eberhart, R.: Parameter selection in particle swarm optimization. In: Proc. 7th Annual Conf. on Evolutionary Programming, pp. 591–600 (1998)
13. Kennedy, J., Eberhart, R.: Swarm Intelligence. Morgan Kaufmann, San Francisco (2001)
14. Poli, R., et al.: Particle swarm optimization: An overview. Swarm Intelligence (August, 2007)
15. Shi, Y., Eberhart, R.: A modified particle swarm optimizer. In: Proc. IEEE Int. Conf. on Evolutionary Computation, pp. 69–73 (1998)
16. Gough, J., et al.: SUPERFAMILY:HMMs representing all proteins of known structure. SCOP sequence searches, alignments, and genome assignments. Nucl. Acids Res. 30(1), 268–272 (2002)
17. Madera, M., et al.: The SUPERFAMILY database in 2004: Additions and improvements. Nucleic Acids Res 32(1), D235–239 (2004)
18. Salwinski, L., et al.: The Database of Interacting Proteins: 2004 update. NAR 32(Database issue), D449–451 (2004)
19. The GNU: Linear Programming Kit (version 4.7), http://www.gnu.org/software/glpk/glpk.html
20. Jansen, R., et al.: A Bayesian Networks Approach for Predicting Protein-Protein Interactions from Genomic Data. Science 302, 449–453 (2003)
21. Rhodes, D.R., et al.: Probabilistic model of the human protein-protein interaction network. Nature Biotechnology 23(8), 951–959 (2005)

Divide, Align and Full-Search for Discovering Conserved Protein Complexes

Pavol Jancura[1,*], Jaap Heringa[2], and Elena Marchiori[1]

[1] ICIS, Radboud University Nijmegen, The Netherlands
[2] IBIVU, Vrije Universiteit Amsterdam, The Netherlands
{jancura,heringa,elena}@few.vu.nl

Abstract. Advances in modern technologies for measuring protein-protein interaction (PPI) has boosted research in PPI networks analysis and comparison. One of the challenging problems in comparative analysis of PPI networks is the comparison of networks across species for discovering conserved modules. Approaches for this task generally merge the considered networks into one new weighted graph, called alignment graph, which describes how interaction between each pair of proteins is preserved in different networks. The problem of finding conserved protein complexes across species is then transformed into the problem of searching the alignment graph for subnetworks whose weights satisfy a given constraint. Because the latter problem is computationally intractable, generally greedy techniques are used. In this paper we propose an alternative approach for this task. First, we use a technique we recently introduced for dividing PPI networks into small subnets which are likely to contain conserved modules. Next, we perform network alignment on pairs of resulting subnets from different species, and apply an exact search algorithm iteratively on each alignment graph, each time changing the constraint based on the weight of the solution found in the previous iteration. Results of experiments show that this method discovers multiple accurate conserved modules, and can be used for refining state-of-the-art algorithms for comparative network analysis.

Keywords: Biological networks alignment, optimization.

1 Introduction

With the recent advances in modern technologies for measuring protein-protein interaction, an exponential increase of data on protein-protein interactions has been generated. Data on thousands of interactions in human and most model species have become available (e.g. [1,2]). Graph-representation of PPI interaction of proteins provides a powerful tool for analyzing and understanding modular organization of cells, for predicting biological functions and for providing insight into a variety of biochemical processes. Recent studies consider a comparative approach for the analysis of PPI networks from different species in order

---

* Corresponding author.

E. Marchiori and J.H. Moore (Eds.): EvoBIO 2008, LNCS 4973, pp. 71–82, 2008.

to discover common protein groups which are likely to share relevant functions [3,4,5]. In particular, this problem is called *pairwise network alignment* when two PPI networks are considered. Algorithms for this problem generally construct a merged weighted graph representation of the two networks, called alignment (or orthology) graph, which describes how interaction between each pair of proteins is preserved in different networks. The problem of finding conserved protein complexes across species is then transformed into the problem of searching the alignment graph for subnetworks whose weights satisfy a given constraint. Due to the computational intractability of such problem, greedy algorithms are commonly used [6,7]. Conserved modules, discovered by computational techniques such as [6], have in general small size compared to the size of the PPI network they belong to. Moreover, as indicated by recent studies, hubs whose removal disconnects the PPI network (articulation hubs) are likely to appear in conserved interaction patterns [8,9]. Based on these motivation, in [10] we introduced an algorithm, called DivA for dividing a pair of PPI networks into small subnets which are expected to cover conserved modules, with the goal of performing modular network alignment. We used this algorithm for performing network alginment in a modular way, by merging pairs of resulting subnets from different species, and then applying an exact optimization algorithm for finding the heaviest subgraph of a weighted graph. Application of this algorithm generates one solution for each alignment subnet. In this paper we propose an extension of this search algorithm which allows to detect an higher number of conserved modules of biological interest. Specifically, the idea is to modify the exact search algorithm for finding the heaviest subgraph of an alignment network, by introducing an upper bound on the maximum weight of the subgraph to be found. Iterated runs of this constrained algorithm are performed, with different values of the upper bound generated at each iteration using the weight of the solution found in the previous iteration. We call this search approach *full-search*. In this way multiple subnets of the alignment network are discovered. The resulting method, called DivAfull, divides each PPI network into subnets using DivA, aligns pairs of subnets from different species, and performs full-search on each aligning pair. We use the state-of-the-art evolution-based alignment graph model introduced in [6] to construct an alignment graph. Results of experiments show effectiveness of the proposed approach, which is capable of detecting an high number of accurate conserved complexes. This number is considerably greater than the number of results identified only by using DivA whereas DivAfull's results contain all DivA's results. Furthermore, we show that improved performance is achieved by merging solutions discovered by DivAfull with those identified by Koyuturk et al.'s algorithm [6].

Recent overviews of approaches and issues in comparative biological networks analysis are presented in [4,5]. Based on the general formulation of network alignment proposed in [3], a number of techniques for (local and global) network alignment have been introduced ([6,7,11,12]). Techniques for local network alignment commonly construct an orthology graph, which provides a merged representation of the given PPI networks, and search for conserved subnets using greedy techniques ([6],[7],[11]). In particular, in [11], d-clusters are defined for searching efficiently between a pair of networks, where a d-cluster consists of d proteins that are close together in the network, and d is a user-given parameter. Another parameter is used for identifying pairs of d-clusters, one from each network, called seeds, which provide starting regions of the alignment graph to be expanded. The algorithm searches for modules conserved across species by expanding these seeds using a greedy technique similar that used in [6],[7]. While the above algorithms focus on network alignment, we focus on 'modular' network alignment. Many papers have investigated the importance of hubs in PPI networks and functional groups [9,13,14,15,16,17]. In particular, it has been shown that hubs with a central role in the network architecture are three times more likely to be essential than proteins with only a small number of links to other proteins [15]. Moreover, if we take functional groups in PPI networks, then, amongst all functional groups, cellular organization proteins have the largest presence in hubs whose removal disconnects the network [9]. Computational techniques for identifying functional modules in PPI networks generally search for clusters of proteins forming dense components [18,19]. The scale-free topology of PPI networks makes difficult to isolate modules hidden inside the central core [20]. In [21] several multi-level graph partitioning algorithms are described addressing the difficulty of partitioning scale-free graphs. The approach we propose differs from the above mentioned works because it does not address (directly) the problem of identifying functional modules in a PPI network, but combines graph-theory, biology and heuristic search for discovering conserved protein complexes in a modular fashion.

2 Divide Align and Full-Search

Given a graph $G = (V, E)$, nodes joined by an edge are called *adjacent*. A *neighbor* of a node u is a node adjacent to u. The degree of u is the number of elements in E containing the vertex u.

Let $G(V, E)$ be a connected undirected graph. A vertex $v \in V$ is called *articulation* if the graph resulting by removing this vertex from G and all its edges, is not connected.

The *Divide algorithm* divides orthologous proteins of the PPI network into subsets. It consists of the following steps:

1. Detect orthologous articulations of the PPI network.
2. Reduce their number by constructing centers using preferential attachment property .
3. In parallel, incrementally expand from each center only alongside orthologous neighbors.
4. Stop when expanding sets are starting to overlap and if they do not have any orthologous neighbor which is not yet added to one of the actual sets.
5. If some orthologous nodes are not in any of the generated set, then join together neighboring ones.

The preferential attachment in the step 2 is a general property of scale-free networks. It means that if a new node is introduced into the network, it will more likely attach to a node of the network with very high degree than to a node with very low degree. Hence, based on this motivation, we construct centers by joining one orthologous articulation hub with its orthologous articulation neighbors, which will more likely to have low degree. The whole algorithm with all technical issues is described in [10].

After dividing, each set of orthologs proteins generates a subnetwork of the PPI network. Pairs of such subnetworks from distinct species can be merged into orthology graphs, which are mined for discovering alignments corresponding to protein complexes conserved across species.

To this aim we use a common approach, based on the construction of a weighted metagraph between two PPI networks of different species. In this metagraph each node corresponds to an homologous pair of proteins, one from each of the two PPI networks. The metagraph is called *alignment* or *orthology graph*. Weights are assigned either to edges, like in [6], or to nodes, like in [7], of the alignment graph using a scoring function. The function transforms conservation and eventually also evolution information to one real value for each edge or node. Induced subgraphs with total weight greater than a given threshold are considered to be relevant *alignments*. In this way one gets two subsets of proteins from each discovered subgraph from the two species, and each such subset provides a conserved complex of proteins.

The problem of finding induced subgraphs with weight greater than a given threshold is reduced in these methods to the problem of finding a maximal induced subgraph. Then an approximation greedy algorithm based on local search is used because the maximum induced subgraph problem is NP-complete (cf. [6]).

In our approach, we align only pairs of subnets from different species having more than one orthologous pair, yielding orthology graphs with more than one node. Because of the small size of the resulting subnets, we use exact optimization [22] for searching in each of such graphs, instead of greedy techniques employed in common approaches.

Specifically, the exact optimization algorithm [22] for finding the maximum weighted induced subgraph is first applied. Then the process is iterated by adding at each iteration the constraint which bounds the weight of the induced subgraph by the weight of the solution found in the previous iteration.

Formally, let f be a function which computes the weight of a subgraph in an input graph and C be a set of constraints which defines an induced subgraph of the input graph. Then we want to maximize the function f on the set defined by constrains C, that is, to solve the following optimization problem:

$$opt = \max_{C} f \qquad\qquad (OptP)$$

Algorithm 1 illustrates the resulting full-search procedure which uses the above constrained optimization problem at each iteration with different bound on the maximum allowed weight.

Algorithm 1. Full Search Algorithm

Input: G: alignment subnetwork, $\varepsilon \geq 0$
Output: List of heavy induced subgraphs of G with weight $> \varepsilon$
1: Formulate the problem of MaxInducedSubGraph for G as $(OptP)$
2: $maxweight = \infty$
3: $C = C + \{opt < maxweight\}$
4: **while** $maxweight > \varepsilon$ **do**
5:    solve $(OptP)$ by an exact method
6:    **if** $opt > \varepsilon$ **then**
7:       record discovered solution
8:    **end if**
9:    $maxweight = opt$
10: **end while**

We call the resulting algorithm `DivAfull`. Finally, redundant alignments are filtered out as done in, e.g., [6]. A subgraph G_1 is said to be *redundant* if there exists another subgraph G_2 which contains $r\%$ of its nodes, where r is a threshold value that determines the extent of allowed overlap between discovered protein complexes. In such a case we say that G_1 *is redundant for* G_2.

3 Evaluation Criteria

In order to assess the performance of our approach, we use the state-of-the-art framework for comparative network analysis proposed in [23], where we change the proposed aligning procedure and searching algorithm to `MaWish` ([8]).

In order to filter out solutions that may also be found when a randomized protein-protein interaction relation between nodes is considered, we apply the following statistical procedure.

1. A collection of 10000 radomized networks are generated by shuffling the edges of the PPI networks while preserving vertex degrees, as well as by shuffling the pairs of homologous proteins while preserving the number of homologous partners per protein.
2. `MaWish` is used for finding solutions on each of the randomized networks.
3. The results are clustered into groups of solutions with equal size (that is, number of subnetwork's nodes). For each size and for each run, the best result (the one with highest score) is recorded. If there is no solution for a given size, we build an artificial cluster consisting of one zero weight solution.
4. For each size, the score at the 95%-percentile, of the corresponding cluster of random solutions, is chosen as treshold for removing 'insignificant' solutions.

We use known *yeast* complexes catalogued in the MIPS database. Category 550, which was obtained from high throughput experiments, is excluded and we retained only manually annotated complexes up to depth 3 in the MIPS tree category structure as standard of truth for quality assessment.

In order to measure statistically significant matches between a solution and a true complex we use the hypergeometric (HG) overlap score. The significance level of a solution is described by means of a function maximizing $-\log(HG)$ through the whole set of true complexes which intersect with the yeast PPI network at least in one protein. Solutions having no annotated protein in the MIPS catalogue are discarded.

We generate again a set of several (10000) radomized networks using the procedure described in the previous section. In each of such networks we find the most significant solution (which maximizes $-\log(HG)$) for each of the considered sizes, by modifying the algorithm MaWish in such a way that it outputs a solution of a given size (number of nodes). Specifically, in the incremental procedure MaWish at each cycle more than one node can be added in order to generate a subgraph with high weight. In the modified version of MaWish we use, if the size of subgraph has reached the given size, we stop. If the size of subgraph has exceeded the given size, we iteratively remove nodes with smallest gain for the actual subgraph, until a subgraph of the given size is obtained.

We compare significance levels of true solutions with those obtained from random networks. In this way we obtain empirical *p-values* for each of the solutions. These *p*-values are further corrected for multiple testing using the false discovery rate (FDR) procedure introduced in [24].

The following notions of specificity, sensitivity and purity are used to assess the quality of the results.

- Let C be the set of solutions with at least one annotated protein in MIPS catalogue and let $C^* \subseteq C$ be the subset of solutions with a significant match ($p < 0.05$). The *specificity* of the solution is defined as $|C^*|/|C|$.
- Let M be the set of true complexes that intersect with the yeast PPI network and let $M^* \subseteq M$ be the subset of complexes with a significant match by a solution. The *sensitivity* of the solution is defined as $|M^*|/|M|$.
- A solution is called *pure* if there exists a true complex whose intersection with the solution covers at least 75% of MIPS annotated proteins in the solution. Let D be the set of all solutions with at least 3 MIPS annotated proteins and let $D^* \subseteq D$ be the subset of pure solutions. The *purity* of the solutions is defined as $|D^*|/|D|$.

4   Results

The two following PPI networks, already compared in [8], are considered: *Saccharomyces cerevisiae* and *Caenorhabditis elegans*, which were obtained from BIND [1] and DIP [2] molecular interaction databases. The corresponding networks consist of 5157 proteins and 18192 interactions, and 3345 proteins and 5988 interactions, respectively. All these data are available at the webpage of MaWish[1]. Moreover, the data already contain the list of potential orthologous

---

[1] *www.cs.purdue.edu/homes/koyuturk/mawish/*

and paralogous pairs, which are derived using BLAST E-values (for more details see [8]). We get 2746 potential orthologous pairs created by 792 proteins in S. cerevisiae and 633 proteins in C. elegans are identified.

We obtain 266 true complexes from the MIPS catalogue whose intersection with the yeast (Saccharomyces cerevisiae) PPI network consist of 876 proteins.

For *Saccharomyces cerevisiae*, 697 articulations, of which 151 orthologs, and 83 centers are identified. After expansion of these centers we covered 639 orthologs. The algorithm assigns the remaining 153 orthologous proteins to 152 new sets.

For *Caenorhabditis elegans*, 586 articulations, of which 158 orthologs, are computed, and 112 centers are constructed from them. Expansion of these centers covers 339 orthologs. The algorithm assigns the remaining orthologous 294 proteins to 288 new sets.

We observe that the last remaining orthologs assigned to new sets without expanding from centers are 'isolated' nodes, in the sense that they are rather distant from each other and not reachable from ortholog paths stemming from centers.

The dividing procedure generates 235 subnets of *Saccharomyces cerevisiae* and 400 subnets of *Caenorhabditis elegans*.

We perform network alignment with MaWish using the same parameter values as those reported in [8]. By constructing alignment graphs between each two subnets from different species containing more than one ortholog pair, we obtain 884 alignment graphs, where the biggest one consists of only 31 nodes.

We apply Algorithm 1 to each of the resulting alignment graphs. Zero weight threshold ($\varepsilon = 0$) is used for considering an induced subgraph as a heavy subgraph or a legal alignment. Redundant graphs are filtered using $r = 80\%$ as the treshold for redundancy.

In this way DivAfull discovers 151 solutions (alignments). By filtering out insignificant results we get 41 solutions.

Using only DivA we get 72 nonredundant alignments against 151 discovered by DivAfull. Because DivA takes only the first best possible solution from each alignment graph, all these solutions are also discovered by DivAfull. This happens in the first iteration of the latter algorithm. In the following iterations, DivAfull discovers other solutions, which have less weight than those discovered in the first iteration. Therefore the best solution can never be filtered out as redundant one. Hence after filtering, DivAfull's results always contain all DivA's solutions and a large number of other, potentially interesting, results identified by applying full search (Algorithm 1).

MaWish yields 83 solutions, and after filtering out insignificant results we get 34 solutions.

For both algorithms, we measure specificity, sensitivity and purity of all solutions and only of significant ones, in order to see whether results consider 'insignificant' are true noise in the data.

Moreover, we compare pairs of redundant alignments as well as new different results. A *paired redundant alignment* is a pair (G_1, G_2) of alignments, with G_1 discovered by DivAfull and G_2 discovered by MaWish, such that either G_1 is

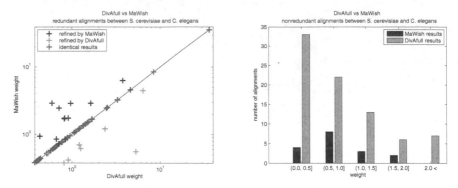

Fig. 1. Analysis all alignments discovered by MaWish and DivAfull. Left figure: Distribution of pairs of weights of paired redundant alignments, one obtained from MaWish and one from DivAfull. Weights of alignments found by DivAfull are on the x-axis, those found by MaWish on the y-axis. '+' is a paired redundant alignment. Right figure: Interval weight distributions of non-redundant alignments discovered by MaWish and DivAfull. The x-axis shows weight intervals, the y-axis the number of alignments in each interval.

redundant for G_2 or vice versa. For a paired redundant alignment (G_1, G_2) we say that G_1 *refines* G_2 if the weight of G_1 is bigger than the weight of G_2.

Results of our experiments are summarized as follows.

Of the 83 solutions of MaWish 56 (67.5%) have at least one MIPS annotated protein and 15 (18.1%) have at least 3 annotated proteins. From the 151 DivAfull results, 103 (68.2%) have at least one annotated protein and 35 (23.2%) have at least 3 annotated proteins.

There are 70 redundant alignments, whose pair of weights are plotted on the left part of Fig. 1. Among these, 48 (31.8% of DivAfull results) are equal (red crosses in the diagonal) and 22(14.6%) different. 8(5.3%) (green crosses below the diagonal) with better DivAfull alignment weight, and 13 (8.6%) (blue crosses above the diagonal) with better MaWish alignment weight (for 1 (0.7%) pair it is undecidable because of rounding errors during computation).

DivAfull finds 81 (53.6%) new alignments, that is, not discovered by MaWish. The right plot of Fig. 1 shows the binned distribution of weights of these alignments, together with the new 17 ones discovered by MaWish but not by DivAfull. There is no significant difference between the overall weight average of the DivAfull (0.8) and the the MaWish (0.86) results.

By considering the union of all alignments discovered by MaWish and DivAfull and by filtering out the redundant ones, 164 alignments are obtained, from which 54.3% consist of refined or new DivAfull ones, and 29.3% consist of alignments discovered by both methods. Of all these alignments 111 have at least one annotated protein and 40 at least with 3 annotated proteins. This results indicate a significant improvement (54.3%) of the performance of MaWish when augmented with DivAfull.

Statistical evaluation of all solution for DivAfull and MaWish, is reported in Table 1. One can observe that DivAfull outperforms MaWish and the number of

Table 1. Specificity, sensitivity and purity for all alignments discovered by `DivAfull` and `MaWish`. The first row of table shows results for combined solutions of both algorithms.

Algorithm	No. of alignments	Specificity (%)	Sensitivity (%)	Purity (%)
DivAfull & MaWish	164	44	6.8	92
DivAfull	151	46	6	91
MaWish	83	43	6	87

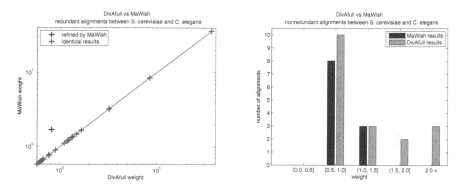

Fig. 2. Analysis significant alignments discovered by `MaWish` and `DivAfull`. Left figure: Distribution of pairs of weights of paired redundant alignments, one obtained from `MaWish` and one from `DivAfull`. Weights of alignments found by `DivAfull` are on the x-axis, those found by `MaWish` on the y-axis. '+' is a paired redundant alignment. Right figure: Interval weight distributions of non-redundant alignments discovered by `MaWish` and `DivAfull`. The x-axis shows weight intervals, the y-axis the number of alignments in each interval.

`DivAfull` solutions is almost double of the number of `MaWish` ones. Combining results obtained by both algorithms generally increases sensitivity and purity, while specificity is increased only w.r.t `MaWish` solutions. The latter phenomenon can be justified by the effect of nonredundant `MaWish` results, since more of them do not have a significant match ($p < 0.05$) and therefore decrease overall specificity when combined with `DivAfull` solutions.

If the same analysis is performed only on the significant alignments then the following results are obtained.

From the significant 34 `MaWish` results, 25 (73.5%) have at least one annotated protein and 4 (11.8%) have at least 3 annotated proteins. From the significant 41 `DivAfull` results, 34 (83%) have at least one annotated protein and 10 (24.4%) have at least 3 annotated proteins.

`DivAfull` finds 18 new alignments not detected by `MaWish`. There are 23 redundant alignments. Among these, 22 (53.7% of `DivAfull` results) are equal and 1 (2.4%) different with better `MaWish` alignment weight.

The right plot of Fig. 2 shows the binned distribution of weights of the 18 (43.9%) found by `DivAfull` but not `MaWish`, and 11 found by `MaWish` and not

Table 2. Specificity, sensitivity and purity for significant alignments discovered by `DivAfull` and `MaWish`. The first row of table shows results for combined significant solutions of both algorithms.

Algorithm	No. of alignments	Specificity (%)	Sensitivity (%)	Purity (%)
DivAfull & MaWish	52	51	4.5	70
DivAfull	41	50	3.4	70
MaWish	34	48	3.8	75

by `DivAfull`. The overall weight average of the `DivAfull` ones (1.609) is greater than the overall average of the `MaWish` ones (0.8536).

By considering the union of all significant alignments of `MaWish` and `DivAfull` and by filtering out the redundant ones, we get together 52 alignments, from which 34.6% results are added as new ones by the `DivAfull` method and 42.3% are equal results discovered by both methods. This shows that performance of the `MaWish` model is improved by 34.6% when the algorithm is augmented with the `DivAfull` method. From all alignments, 41 have at least one annotated protein and 10 at least with 3 annotated proteins.

Table 2 report statistical evaluation of results of `MaWish`, `DivAfull`, and their union. `DivAfull` solutions have better specificity than `MaWish` solutions and similar sensitivity. Concerning purity, `DivAfull` has 7 pure solutions from 10 considered, while `MaWish` has 3 pure solutions from 4. Because of the small number of the considered alignments, the purity measure in this case does not provide sufficient information for comparing the two algorithms. Considering the union of `MaWish` and `DivAfull` generally increases sensitivity and specificity. Moreover, the new solutions added by `DivAfull` increase the number of pure alignments.

In summary, these results show that `DivAfull` can be successfully applied to discover conserved protein complexes and to 'refine' state-of-the-art algorithms for network alignment.

5   Conclusion

This paper introduced a heuristic algorithm, `DivAfull`, for discovering conserved protein complexes, which is an extension of a previously proposed algorithm, `DivA`. Results of the comparative experimental analysis indicated that `DivAfull` improves the search procedure of `DivA`. Moreover, comparison between `MaWish` and `DivAfull` indicated that `DivAfull` is able to discover new alignments which significantly increase the number of discovered complexes. `DivAfull` solutions showed also improved match with well-know yeast complexes measured by specificity, sensitivity and purity. Combination of solutions discovered by both `MaWish` and `DivAfull`, yielded new and refined alignments.

Although using an exact search in `DivAfull` requires higher computational time than the greedy searching of `MaWish` (in our experiment it took more than 4 hours on a desktop machine (AMD Athlon 64 Processor 3500+, 2 GB RAM)), the advantage of a modular approach relies also in possible parallelization of parts of the method. For instance, the full search algorithm can be run independently on each alignment graph. Moreover, ad-hoc internal parallelization can be applied to improve efficiency. We are actually working on such optimized implementation of `DivAfull`.

Results show that the filtering procedure used for removing 'insignificant' results seems to be rather strict, because it appears to discard a substantial number of solutions which seem to be biologically meaningful. A more thorough analysis of real biological functions of some of the new discovered results is still needed.

Finally, we intend to analyze instances of our approach based on other methods, such as [7].

References

1. Bader, G.D., Donaldson, I., Wolting, C., Ouellette, B.F.F., Pawson, T., Hogue, C.W.V.: Bind–the biomolecular interaction network database. Nucleic Acids Res 29(1), 242–245 (2001)
2. Xenarios, I., Salwínski, Ł., Duan, X.J., Higney, P., Kim, S.M., Eisenberg, D.: Dip, the database of interacting proteins: A research tool for studying cellular networks of protein interactions. Nucleic Acids Research 30(1), 303–305 (2002)
3. Kelley, B.P., Sharan, R., Karp, R.M., Sittler, T., Root, D.E., Stockwell, B.R., Ideker, T.: Conserved pathways within bacteria and yeast as revealed by global protein network alignment. Proceedings of the National Academy of Science 100, 11394–11399 (2003)
4. Sharan, R., Ideker, T.: Modeling cellular machinery through biological network comparison. Nature Biotechnology 24(4), 427–433 (2006)
5. Srinivasan, B.S., Shah, N.H., Flannick, J., Abeliuk, E., Novak, A., Batzoglou, S.: Current Progress in Network Research: Toward Reference Networks for kKey Model Organisms. Brief. in Bioinformatics (Advance access) (2007)
6. Koyutürk, M., Grama, A., Szpankowski, W.: Pairwise local alignment of protein interaction networks guided by models of evolution. In: Miyano, S., Mesirov, J., Kasif, S., Istrail, S., Pevzner, P.A., Waterman, M. (eds.) RECOMB 2005. LNCS (LNBI), vol. 3500, pp. 48–65. Springer, Heidelberg (2005)
7. Sharan, R., Ideker, T., Kelley, B.P., Shamir, R., Karp, R.M.: Identification of protein complexes by comparative analysis of yeast and bacterial protein interaction data. Journal of Computational Biology 12(6), 835–846 (2005)
8. Koyutürk, M., Kim, Y., Topkara, U., Subramaniam, S., Grama, A., Szpankowski, W.: Pairwise alignment of protein interaction networks. Journal of Computational Biology 13(2), 182–199 (2006)
9. Pržulj, N.: Knowledge Discovery in Proteomics: Graph Theory Analysis of Protein-Protein Interactions. CRC Press, Boca Raton (2005)
10. Jancura, P., Heringa, J., Marchiori, E.: Dividing protein interaction networks by growing orthologous articulations. IR-BIO-002 (2007), http://www.cs.vu.nl/~elena/diva.pdf
11. Flannick, J., Novak, A., Srinivasan, B.S., McAdams, H.H., Batzoglou, S.: Graemlin: General and robust alignment of multiple large interaction networks. Genome Res. 16(9), 1169–1181 (2006)

12. Singh, R., Xu, J., Berger, B.: Global Alignment of Multiple Protein Interaction Networks. In: Speed, T., Huang, H. (eds.) RECOMB 2007. LNCS (LNBI), vol. 4453, Springer, Heidelberg (2007)
13. Pržulj, N., Wigle, D., Jurisica, I.: Functional topology in a network of protein interactions. Bioinformatics 20(3), 340–384 (2004)
14. Rathod, A.J., Fukami, C.: Mathematical properties of networks of protein interactions. CS374 Fall 2005 Lecture 9. Computer Science Department, Stanford University (2005)
15. Jeong, H., Mason, S.P., Barabasi, A.-L., Oltvai, Z.N.: Lethality and centrality in protein networks. NATURE v 411, 41 (2001)
16. Ekman, D., Light, S., Björklund, A.K., Elofsson, A.: What properties characterize the hub proteins of the protein-protein interaction network of saccharomyces cerevisiae? Genome Biology 7(6), R45 (2006)
17. Ucar, D., Asur, S., Catalyurek, U., Parthasarathy, S.: Improving functional modularity in protein-protein interactions graphs using hub-induced subgraphs. In: Fürnkranz, J., Scheffer, T., Spiliopoulou, M. (eds.) PKDD 2006. LNCS (LNAI), vol. 4213, Springer, Heidelberg (2006)
18. Bader, G.D., Lässig, M., Wagner, A.: Structure and evolution of protein interaction networks: A statistical model for link dynamics and gene duplications. BMC Evolutionary Biology 4(51) (2004)
19. Li, X.L., Tan, S.H., Foo, C.S., Ng, S.K.: Interaction graph mining for protein complexes using local clique merging. Genome Informatics 16(2), 260–269 (2005)
20. Yook, S.H., Oltvai, Z.N., Barabási, A.L.: Functional and topological characterization of protein interaction networks. PROTEOMICS 4, 928–942 (2004)
21. Abou-Rjeili, A., Karypis, G.: Multilevel algorithms for partitioning power-law graphs. In: 20th International Parallel and Distributed Processing Symposium (IPDPS) (2006)
22. Wolsey, L.A.: Integer Programming, 1st edn. Wiley-Interscience (September 9, 1998)
23. Hirsh, E., Sharan, R.: Identification of conserved protein complexes based on a model of protein network evolution. Bioinformatics 23(2), e170–176 (2007)
24. Benjamini, Y., Hochberg, Y.: Controlling the false discovery rate: A practical and powerful approach to multiple testing. Journal of the Royal Statistical Society. Series B (Methodological) 57(1), 289–300 (1995)

Detection of Quantitative Trait Associated Genes Using Cluster Analysis

Zhenyu Jia[1,*], Sha Tang[2], Dan Mercola[1], and Shizhong Xu[3]

[1] Department of Pathology and Laboratory Medicine, University of California, Irvine, CA 92697, USA
{zjia,dmercola}@uci.edu
[2] Department of Pediatrics, University of California, Irvine, CA 92697, USA
shat@uci.edu
[3] Department of Botany and Plant Sciences, University of California, Riverside, Riverside, CA 92521, USA
shxu@ucr.edu

Abstract. Many efforts have been involved in association study of quantitative phenotypes and expressed genes. The key issue is how to efficiently identify phenotype-associated genes using appropriate methods. The limitations for the existing approaches are discussed. We propose a hierarchical mixture model in which the relationship between gene expressions and phenotypic values is described using orthogonal polynomials. Gene specific coefficient, which reflects the strength of association, is assumed to be sampled from a mixture of two normal distributions. The association status for a gene is determined based on which distribution the gene specific coefficient is sampled from. The statistical inferences are made via the posterior mean drawn from a Markov Chain Monte Carlo sample. The new method outperforms the existing methods in simulated study as well as the analysis of a mice data generated for obesity research.

Keywords: Gibbs sampler, Microarray.

1 Introduction

Microarray technology allows us to measure the expression levels of many thousands of genes simultaneously. The objective of microarray experiments is to closely examine the changes of gene expression under different experimental conditions. These conditions may simply be control and various treatments [1], or may represent different time slots after a certain treatment is applied to the experimental subjects [2], or may refer to measurements of quantitative phenotype for different subjects [3]. Many statistical methodologies have been proposed to analyze data generated from microarray experiments. Fundamental microarray data analyses aim to identify a list of genes as being differentially expressed across experimental conditions [4,5,6]. Recent methods, such as various cluster analyses, have been devised not to find individual genes but to search for groups

---

* To whom all correspondence should be addressed.

E. Marchiori and J.H. Moore (Eds.): EvoBIO 2008, LNCS 4973, pp. 83–94, 2008.

of genes that are functionally related [7,8,9,10]. However, many existing cluster analyses only use expression data, which requires extra steps to infer the functions of genes that form a cluster. Incorporating biological information [11] or phenotypic information [12] into cluster analyses seems to be more efficient and reasonable.

Efforts have been provided to uncover genes that affect the phenotype of interest. For example, [3] conducted an experiment to study the relationship of gene expression and Alzheimer's disease. The Pearson's correlation was calculated for each gene with the phenotypic values separately. Genes were declared to be disease-associated if their correlation coefficients are statistically significant. Similarly, [13] used the Pearson's correlation analysis to study the relationship of gene expression and mouse weight. First, the highly correlated genes were clustered into the same group which was called "module". Next, they assessed the physiological relevance of each module by examining the overall correlation of the module genes with the phenotype. The genes within a significant module were claimed to be associated with the phenotype.

As suggested by [12], the Pearson's correlation analysis may not be optimal for two reasons: (1) Genes are not jointly analyzed leading to a poor information sharing across genes. (2) A significant correlation is not always biologically meaningful unless the regression is also high. They proposed a mixed model in which the gene expression levels are linearly regressed on the phenotypic values. The regression coefficients, which reflect the affiliation of the genes with the phenotype, are used to cluster genes into a number of functional groups. A cluster is claimed to be significant if the regression coefficient of the mean expression profile is not equal to zero; otherwise, the cluster is claimed as neutral. Genes that have been assigned into non-neutral clusters are target genes. Because the mixed model of [12] is solely built upon the assumption of linear association, it is limited to pick up genes that are associated with the phenotype in a non-linear manner. In order to solve this problem, [14] developed a mixed model to cluster genes based on the non-linear association using orthogonal polynomials. For these two model-based analyses, the optimal number of clusters is not known and needs to be determined by comparing the BIC [15] values for different models. To our experience, this often requires credible evaluations for at least 10 models with distinct dimensionality of parameters which is defined by the number of clusters. It would be more computationally intensive if non-linear association is considered due to the complex nature of microarray data.

In current study, we developed a Bayesian hierarchical model to cluster genes with fixed number of clusters. The non-linear relationship between gene and the phenotype is also described using orthogonal polynomials. The orthogonal polynomials can be constructed as described by [16]. For each gene, the coefficient of each polynomial is assumed to be sampled from a two-components mixture Normal distribution. Both Normal components have mean zero but different variances, i.e., one has a very small variance while the other has a larger variance. If the corresponding coefficient is sampled from the component with small variance, the coefficient is enforced to be zero; otherwise, the coefficient is non-trivial and its magnitude should be estimated from data. That is to say each gene may be assigned into one of two clusters based on whether it is associated with the polynomial. Suppose that there are p polynomials in the expression model. Therefore, there are a total of 2^p patterns or clusters to illustrate genes under study. Once p is chosen, the number of clusters is immediately determined, which circumvents the model evaluations required by the aforementioned methods.

2   Methods

2.1   Hierarchical Linear Model

Let m and N be the number of genes and the number of subjects under study, respectively. Let Z be the measurements of a quantitative phenotype collected from N subjects. The expression levels of gene i across N subjects can be described in the following model:

$$Y_i(Z) = \alpha_i + \beta_i(Z) + \epsilon_i, \tag{1}$$

where $i = 1, \ldots, m$. In the model 1, $Y_i(Z)$ is a $N \times 1$ matrix, α_i represents the gene specific intercept, $\beta_i(Z)$ is an arbitrary function chosen to describe the relationship between the gene expressions and the phenotypic values, and ϵ_i is used to model the random error with assumed $N(\mathbf{0}, I\sigma^2)$ distribution.

There are different ways to choose function $\beta_i(Z)$. In current study, we only consider the orthogonal polynomials [16], such that, $\beta_i(Z)$ can be expressed as:

$$\beta_i(Z) = X\beta_i = \sum_{j=1}^{p} X_j \beta_{ij},$$

where p is the degree of orthogonal polynomials after transformation, Z is transformed into a $N \times p$ matrix which is denoted by $X = (X_1, \ldots, X_p)$, and $\beta_i = (\beta_{i1}, \ldots, \beta_{ip})$ represents the corresponding coefficients for gene i. Then, model 1 can be rewritten as:

$$Y_i = \alpha_i + \sum_{j=1}^{p} X_j \beta_{ij} + \epsilon_i. \tag{2}$$

Using a linear contrasting scheme (see [14]), model 2 can be further written as

$$y_i = \sum_{j=1}^{p} x_j \beta_{ij} + \varepsilon_i, \tag{3}$$

where $\sum_{k=1}^{N} y_{ik} = 0$ and $\sum_{k=1}^{N} x_{jk} = 0$. In fact, we do not have N pieces of independent information for each gene after linear contrasting. Therefore, the last element of vector y_i should be removed and y_i becomes an $n \times 1$ vector for $n = N - 1$. Accordingly, x_j becomes an $n \times 1$ vector, and ε_i is now $N(\mathbf{0}, R\sigma^2)$ distributed with a known $n \times n$ positive definite matrix R (see [14]).

For the sake of convenience for presentation, we use the following notation to express different probability densities throughout current study:

$$p(\text{variable}|\text{parameter list}) = \text{DensityName}(\text{variable}; \text{parameter list}).$$

For example, the probability of y_i given all the β_{ij} variables and σ^2 is described as

$$p\left(y_i \left| \sum_{j=1}^{p} x_j \beta_{ij}, R\sigma^2 \right.\right) = \text{Normal}\left(y_i; \sum_{j=1}^{p} x_j \beta_{ij}, R\sigma^2\right). \tag{4}$$

The model 3 is the lowest level in the hierarchical structure, which is governed by higher parameters, such as regression coefficients (β_{ij}) and the residual variance (σ^2). These parameters themselves are controlled by assumed higher distributions. In this study, we assign a mixture distribution to β_{ij} as originally suggested by [17],

$$p(\beta_{ij}|\eta_{ij}, \sigma_j^2) = (1 - \eta_{ij})\text{Normal}(\beta_{ij}; 0, \delta) + \eta_{ij}\text{Normal}(\beta_{ij}; 0, \sigma_j^2) \tag{5}$$

where $\delta = 10^{-4}$ (a small positive number) and σ_j^2 is an unknown variance assigned to the jth polynomial. Variable $\eta_{ij} = \{0, 1\}$ is used to indicate whether β_{ij} is sampled from a $N(0, \delta)$ or a $N(0, \sigma_j^2)$ distribution. If it comes from the first normal distribution, β_{ij} is virtually fixed at zero; otherwise, β_{ij} has a non-trivial value and should be estimated from the data. Therefore, $\eta_{ij} = 1$ means that $\beta_{ij} \neq 0$ and gene i is associated with the jth polynomial. The hierarchical level of density 5 is regulated by η_{ij} and σ_j^2. We further describe η_{ij} by a Bernoulii distribution with probability ρ_j, denoted by

$$p(\eta_{ij}|\rho_j) = \text{Bernoulii}(\eta_{ij}; \rho_j). \tag{6}$$

The parameter ρ_j will control the proportion of the genes that are associated with the jth polynomial. Because of the hierarchical nature, we may further describe ρ_j by a Dirichlet distribution, denoted by $\text{Dirichlet}(\rho_j; 1, 1)$. The variance components of the hierarchical model are assigned scaled inverse chi-square distributions, denoted by $\text{Inv} - \chi^2(\sigma_j^2; d_0, \omega_0)$. We choose $d_0 = 5$ and $\omega_0 = 50$ for σ_j^2, and choose $d_0 = 0$ and $\omega_0 = 0$ for σ^2.

2.2   Markov Chain Monte Carlo

The typical technique for inferring the posterior distributions of the parameters is to use MCMC sampling since the posterior distributions are intractable. We draw a posterior sample from which empirical posterior means of interested parameters can be found. First, we choose initial values for parameter θ, where $\theta = (\sigma^2, \sigma_1^2, \ldots, \sigma_p^2, \rho)$. We then derive the distribution of one parameter conditional on the data and values of all other variables, i.e., $p(\theta_k|\text{data}, \theta_{-k})$, where θ_k is current parameter of interest and θ_{-k} is the list of remaining variables. This distribution usually has a simple form from which a value for θ_k can be sampled.

The parameter θ_k is then updated using the realized value, and it will be used as known parameter to update all other parameters in the same manner. The detailed sampling scheme for each variable is described as follows.

(1) Variable η_{ij} is simulated from Bernoulli($\eta_{ij}; \pi_{ij}$), where

$$\pi_{ij} = \frac{\rho_j N(\gamma_{ij}; 0, \sigma_j^2)}{\rho_j N(\gamma_{ij}; 0, \sigma_j^2) + (1 - \rho_j) N(\gamma_{ij}; 0, \delta)} \tag{7}$$

(2) Variable β_{ij} is simulated from $N(\beta_{ij}; \mu_\beta, \sigma_\beta^2)$, where

$$\mu_\beta = \left[x_j^T R^{-1} x_j + \frac{\sigma^2}{\eta_{ij} \sigma_j^2 + (1 - \eta_{ij}) \delta} \right]^{-1} x_j^T R^{-1} \triangle y_i, \tag{8}$$

$$\sigma_\beta^2 = \left[x_j^T R^{-1} x_j + \frac{\sigma^2}{\eta_{ij} \sigma_j^2 + (1 - \eta_{ij}) \delta} \right]^{-1} \sigma^2 \tag{9}$$

and

$$\triangle y_i = y_i - \sum_{j' \neq j}^{p} x_{j'} \beta_{ij'} \tag{10}$$

which is called the offset of y_i adjusted for the jth polynomial effect.

(3) Sample σ_j^2 from

$$\text{Inv} - \chi^2 \left(\sigma_j^2; \sum_{i=1}^{m} \eta_{ij} + 5, \sum_{i=1}^{m} \eta_{ij} \beta_{ij}^2 + 50 \right).$$

(4) Sample σ^2 from

$$\text{Inv} - \chi^2 \left(\sigma^2; mn, \sum_{i=1}^{m} (y_i - \sum_{j=1}^{p} x_j \beta_{ij})^T R^{-1} (y_i - \sum_{j=1}^{p} x_j \beta_{ij}) \right).$$

(5) Simulate ρ_j from

$$\text{Dirichlet} \left(\rho_j; \sum_{i=1}^{m} \eta_{ij} + 1, m - \sum_{i=1}^{m} \eta_{ij} + 1 \right).$$

So far, every variable has been updated. Once every variable is updated, we complete one iteration or sweep. The sampling process continues until the Markov chain reaches its stationary distribution. The length of the chain required for convergence can be determined by the R package "coda" [18]. We discard a number of iterations from the beginning of the chain, which is so-called burn-in period. For the remaining portion of the chain, we save one observation in every 10 sweeps to form a posterior sample until the sample is sufficiently large to allow an accurate estimate of the posterior mean for each variable. Let

$\bar{\eta}_{ij} = N_p^{-1} \sum_{l=1}^{N_p} \eta_{ij}^{(l)}$ be the posterior mean of variable η_{ij}, where N_p is the posterior sample size. Gene i is said to be associated with the jth polynomial if $\bar{\eta}_{ij}$ is greater than some pre-specified threshold. We use 0.8 as such cutoff point throughout the current study since [19] showed that 0.8 was quite sufficient to achieve the false discovery rate (FDR) control at $\leq 1\%$ level in the similar analysis.

3   Implementation

3.1   Simulation Study

In the simulation study, a total of 20 datasets were simulated independently. For each dataset, expression levels of 1000 genes were simulated for 50 subjects. The phenotypic value for each subject was randomly selected from U(0, 10). The 50×1 phenotype matrix was then transformed into 50×3 orthogonal polynomials matrix with degree 3. The corresponding 3×1 regression coefficient matrix for each gene was generated as follows. For genes 1 to 5 and genes 21 to 35, the coefficients for the polynomial of the first order were simulated from $N(0, 3^2)$. For genes 6 to 10, genes 16 to 25, and genes 31 to 35, the coefficients for the polynomial of the second order were simulated from $N(0, 1^2)$. For genes 11 to 20 and genes 26 to 35, the coefficients for the polynomial of the third order were simulated from $N(0, 0.5^2)$. In current study, we define a gene-polynomial association as a linkage. Thus, a total of 60 linkages were generated in the simulation study. Such set up made the 1000 genes fall into $2^3 = 8$ binary-based categories, which were represented by (0 0 0), (1 0 0), (0 1 0), (0 0 1), (0 1 1), (1 1 0), (1 0 1) and (1 1 1), respectively. For example, gene 1 can be regarded as a member of the cluster (1 0 0) since it was only associated with the polynomial of the first order; while gene 35 belonged to the cluster (1 1 1) because the coefficients for all three polynomials are non-trivial. Only the first 35 genes were associated with phenotype while the majority of the genes were placed in the neutral cluster represented by (0 0 0). The residual error for each gene was sampled from $N(0, 0.4^2)$. The aim of our analysis is to detect genes represented by significant linkages with the phenotypic polynomials.

We used the new method to analyze the 20 simulated datasets separately. The results summarized from 20 analyses are presented in Table 1. The estimated parameters agreed with the true values very well. Due to the small sample size

Table 1. True and estimated values by the new method for the parameters used in simulation study

	\multicolumn{7}{c}{Parameter}						
	ρ_1	ρ_2	ρ_3	σ_1^2	σ_1^2	σ_1^2	σ^2
True	0.020	0.020	0.020	9.00	1.00	0.25	0.160
Estimate	0.021	0.019	0.017	10.72	3.18	2.93	0.160

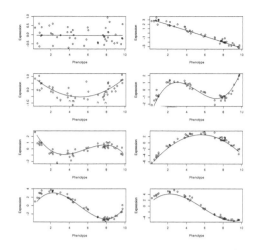

Fig. 1. Plots of the typical genes selected from each of 8 clusters for the analysis of one simulated dataset

Table 2. Comparison between two methods based on the percentages of true genes identified by each of them. Dataset one is the one simulated in current study and dataset two is that simulated in [14]. Method I is the proposed method and method II is the method of [14].

	Dataset	
Method	One	Two
I	88.57%	100.0%
II	62.86%	98.38%

(≤ 30) for each polynomial, the estimated σ_j^2 showed some deviations from the true values. Figure 1 gives the plots of the typical genes selected from each of 8 clusters for the analysis of one simulated dataset. The expression pattern of each gene across phenotypic values is satisfactorily depicted with a regression curve approximated by the new method. We also used the method of [14] to analyze the same dataset. The optimal BIC occurred when the number of clusters was set to 7. Because two methods use different criterions to cluster genes, that is the method of [14] classifies genes based on their mean expression pattern across the phenotypic polynomials; while the proposed method clusters genes based on whether they are significantly associated with the phenotypic polynomials. Thus, we compared two methods by checking the proportions of true associated genes that have been successfully identified by each of them. Note that the numbers of falsely identified genes were zero for both methods. The results are listed in Table 2, from where we can see that the new method identified more true genes than the method of [14]. We examined all 7 true linkages missed by our analysis. The average of the absolute effects for these 7 linkages was 0.08, which is too small to be detectable with reasonable analysis methods. We understand that the better performance of the new method may result from simulation scheme which could be biased to the new method. To eliminate this nuisance factor, we also analyzed the dataset simulated in [14] using two methods. From Table 2, we can see that the new method outperformed the method of [14] again.

3.2   Analysis of Mice Data

To demonstrate the new method, we analyzed a mice data collected for obesity study by [20]. The data are publicly available at gene expression omnibus (GEO)

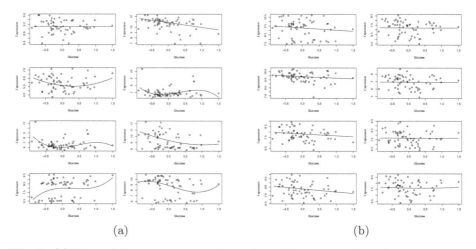

(a)   (b)

Fig. 2. (a) Plots of the typical transcripts selected from each of the 8 clusters for the analysis of glucose-expression associations for mice data. (b) Scatter plots of selected transcripts identified by one method by missed by the other method for the analysis of glucose-expression associations for mice data. Four transcripts on the left panel are those only detected by the proposed method; while the other four on the right panel are transcripts solely detected by the method of [14].

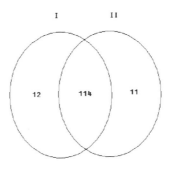

Fig. 3. Transcripts detected by the two methods for the analysis of glucose-expression associations for mice data. Method I is the proposed method and method II is the method of [14].

with accession no. GSE3330. In their experiment, a total of 60 ob/ob mice were examined. For each mouse, the expression levels of over 40,000 transcripts and 25 obesity related phenotypes were measured. Since the expression levels of most transcripts are constant across 60 mice and they do not provide any information, we eliminated those non-variant transcripts prior to the analysis to lessen the computation burden. We sorted all transcripts by their variances across 60 individuals and deleted the transcripts with variances less than 0.05, leaving 5185 most varying transcripts for further analysis. Similar pre-screening scheme has been used for array data analyses [13,19]. In current study, we only investigated the association between gene expression and plasma glucose level (mg/dl). The phenotypic data were collected at eight weeks of age. A total of 126 transcripts were detected to be associated with the glucose level. The typical transcripts selected from each of the 8 clusters are presented in Figure 2(a). We also used the method of [14] to analyze the expression-glucose data. The optimal BIC occurred when the number of clusters was 2, which might not be sound. More distinct clusters were expected due to the complexity of array data. The BIC value kept going down as the number of clusters increased, though the differences of analytic results from different models were trivial. In this case we chose the number of clusters as 8 to achieve the parallel between two methods. A total of 125 transcripts were identified. From Figure 3, we can see that both methods detected 114 common transcripts. We checked all the transcripts that have been detected by one method but missed by the other one. The left four transcripts in Figure 2(b) were detected by the new method but missed by the method of [14]. These four transcripts had slight slopes which should be accounted. The four transcripts on the right panel were only detected by the method of [14]. We could not see any regressions between the expression levels and the phenotypic values. It seemed that the new method had more power than the method of [14]; on the other hand, the new method was subject to lower type I error than the method of [14].

4   Discussion

The purpose of current study is to introduce a more sensitive and convenient approach for association study on gene expressions and quantitative traits. Similar to the existing method of [14], the new method is also based on the non-linear relationship assumption and is realized via orthogonal polynomial transformation. The differences are: (1) the method of [14] organize genes based on their mean expression patterns across phenotypic values; while the new method clusters genes by examine their associations with the polynomials of phenotype. (2) in method of [14], extra model evaluations are needed to find the optimal number of clusters; however, this is not necessary for new method, where the number of clusters is always fixed. In the analysis of [14], the significant tests are performed on the coefficients of the mean expressions for genes. In such the case, the coefficients for all polynomials are jointly considered, which sometimes leads to loss of power. Suppose that, for a gene, the coefficients for the second and the third order are trivial, while the coefficient for the first order is somewhat significant. This gene may not be detected when the overall significance is considered. In the new method, this would not happen since we test the association of gene expression with each polynomial individually. We can sharpen the prior (δ) to make the analysis sensitive to a satisfactory extent. A gene that is significantly linked to any of the polynomials will be picked up. We also noticed that all the missed significant genes by the method of [14] are all linearly associated with the phenotype, which means that the genes linked with phenotype with higher orders are relatively easier to be seen. That makes senses because the high-order association tends to show a more obvious pattern than the linear association does. This explains why the genes with slight first order regressions have been overlooked by the method of [14]. For the analysis of [14], we need to compare the BIC values for different models to find the optimal number of clusters, which requires considerable extra effort. Such evaluation failed probably due to the complexity of the array data. We consider this extra computation can be avoided by fixing the number of clusters through a meaningful way. In current study, we classify genes into one of two clusters for each polynomial, that is, cluster contains non-associated genes and cluster contains associated genes. Thus, the association of a gene with the phenotype may be describe by one of 2^p patterns, which makes the new method more efficient in implementation.

As aforementioned in Methods section, different functions can be adopted for $\beta_i(Z)$, which is used to describe the relationship between the gene expression and the phenotype in current study. For example, we may use B-spline transformation instead. Simulation studies indicated that B-spline version is equivalent to orthogonal polynomials version (data not shown). B-spline is a alternative way of constructing a basis for piecewise polynomial; however, it is not a natural method of describing spline. Thus, we prefer using orthogonal polynomials version in current study since the behavior of regression of gene expressions on phenotype can be easily interpreted.

The association study of gene expressions and phenotypes provides with a pilot research for gene network study. For example, we may first identify transcripts that are associated with the phenotypes of the disease. The common genes that have been discovered to be associated with multiple phenotypes may play key roles in disease development. Experiments may be carried out to verify their biological significance. We may treat the validated genes as seed genes and further search for other non-annotated genes that may be functionally connected with these genes. New genes may be identified by checking if their expression levels significantly correlate with that of the seed genes. Or, given the information on the genomic markers, we may map the seed genes as well as the genes with unknown functions jointly using so called eQTL mapping scheme, such as [21,19]. The genes that have been mapped to the same genomic loci with the seed genes are likely to be functionally related to the seed genes and contribute to the disease.

Acknowledgment

This research was supported by the National Institute of Health Grant R01-GM55321 and the National Science Foundation Grant DBI-0345205 to SX, and by the National Institute of Health SPECS Consortium Grant CA 114810-02.

References

1. Dudoit, S., Yang, Y.H., Callow, M.J., Speed, T.P.: Statistical methods for identifying differentially expressed genes in replicated cdna microarray experiments. Stat. Sinic. 12, 111–139 (2002)
2. Saban, M.R., Hellmich, H., Nguyen, N.B., Winston, J., Hammond, T.G., Saban, R.: Time course of lps- induced gene expression in a mouse model of genitourinary inflammation. Physiological Genomics 5, 147–160 (2001)
3. Blalock, E.M., Geddes, J.W., Chen, K.C., Porter, N.M., Markesbery, W.R., Landfield, P.W.: Incipient alzheimer's disease: Microarray correlation analyses reveal major transcriptional and tumor suppressor responses. Proceedings of the National Academy of Sciences of the United States of America 101, 2173–2178 (2004)
4. Tusher, V.G., Tibshirani, R., Chu, G.: Significance analysis of microarrays applied to the ionizing radiation response. Proceedings of the National Academy of Sciences of the United States of America 98, 5116–5121 (2001)
5. Efron, B., Tibshirani, R., Storey, J.D., Tusher, V.: Empirical bayes analysis of a microarray experiment. J. Am. Stat. Assoc. 96, 1151–1160 (2001)
6. Newton, M.A., Noueiry, A., Sarkar, D., Ahlquist, P.: Detecting differential gene expression with a semiparametric hierarchical mixture method. Biostatistics 5, 155–176 (2004)
7. Brown, M.P.S., Grundy, W.N., Lin, D., Cristianini, N., Sugnet, C.W., Furey, T.S., Ares, M., Haussler, D.: Knowledge-based analysis of microarray gene expression data by using support vector machines. Proceedings of the National Academy of Sciences of the United States of America 97, 262–267 (2000)
8. Herrero, J., Valencia, A., Dopazo, J.: A hierarchical unsupervised growing neural network for clustering gene expression patterns. Bioinformatics 17, 126–136 (2001)
9. Yeung, K.Y., Fraley, C., Murua, A., Raftery, A.E., Ruzzo, W.L.: Model-based clustering and data transformations for gene expression data. Bioinformatics 17, 977–987 (2001)
10. Lazzeroni, L., Owen, A.: Plaid models for gene expression data. Statistica Sinica 12, 61–86 (2002)
11. Huang, D.S., Pan, W.: Incorporating biological knowledge into distance-based clustering analysis of microarray gene expression data. Bioinformatics 22, 1259–1268 (2006)
12. Jia, Z., Xu, S.: Clustering expressed genes on the basis of their association with a quantitative phenotype. Genetical Research 86, 193–207 (2005)
13. Ghazalpour, A., Doss, S., Zhang, B., Wang, S., Plaisier, C., Castellanos, R., Brozell, A., Schadt, E.E., Drake, T.A., Lusis, A.J., Horvath, S.: Integrating genetic and network analysis to characterize genes related to mouse weight. Plos Genetics 2, 1182–1192 (2006)
14. Qu, Y., Xu, S.H.: Quantitative trait associated microarray gene expression data analysis. Molecular Biology and Evolution 23, 1558–1573 (2006)
15. Schwartz, G.: Estimating the dimensions of a model. Ann. Stat. 6, 461–464 (1978)

16. Hayes, J.G.: Numerical methods for curve and surface fitting. J. Inst. Math. Appl. 10, 144–152 (1974)
17. George, E.I., Mcculloch, R.E.: Variable selection via gibbs sampling. Journal of the American Statistical Association 88, 881–889 (1993)
18. Raftery, A.E., Lewis, S.M.: One long run with diagnostics: Implementation strategies for markov chain monte carlo. Statistical Science 7, 493–497 (1992)
19. Jia, Z., Xu, S.: Mapping quantitative trait loci for expression abundance. Genetics 176, 611–623 (2007)
20. Lan, H., Chen, M., Flowers, J.B., Yandell, B.S., Stapleton, D.S., Mata, C.M., Mui, E.T., Flowers, M.T., Schueler, K.L., Manly, K.F., Williams, R.W., Kendziorski, C., Attie, A.D.: Combined expression trait correlations and expression quantitative trait locus mapping. Plos Genetics 2, e6 (2006)
21. Schadt, E.E., Monks, S.A., Drake, T.A., Lusis, A.J., Che, N., Colinayo, V., Ruff, T.G., Milligan, S.B., Lamb, J.R., Cavet, G., Linsley, P.S., Mao, M., Stoughton, R.B., Friend, S.H.: Genetics of gene expression surveyed in maize, mouse and man. Nature 422, 297–302 (2003)

Frequent Subsplit Representation of Leaf-Labelled Trees*

Jakub Koperwas and Krzysztof Walczak

Institute of Computer Science, Warsaw University of Technology,
Nowowiejska 15/19, 00-665 Warsaw, Poland
J.Koperwas@elka.pw.edu.pl, K.Walczak@ii.pw.edu.pl

Abstract. In this paper we propose an innovative method of representing common knowledge in leaf-labelled trees as a set of frequent subsplits, together with its interpretation. Our technique is suitable for trees built on the same leafset as well as for trees where the leafset varies. The proposed solution has a very good interpretation, as it returns different, maximal sets of taxa that are connected with the same relations in the input trees. In contrast to other methods known in literature it does not necessarily result in one tree, but may result in a profile of trees, which are usually more resolved than the consensus trees.

Keywords: Leaf-labelled trees, phylogenetic trees, frequent subsplits.

1 Introduction

Various data from different fields of science are represented as structured or semi-structured data, for example trees. Therefore, tree mining techniques are worth studying. Among others, we can distinguish leaf-labelled trees as a separate type of trees. Leaf-labelled trees play an important role in bioinformatics, as they represent phylogenetic and tandem duplication trees. One of the key issues in leaf-labelled trees analysis is common information representation [1,2]. The existing techniques have several disadvantages, among others they are not applicable to trees where the leafset varies (we will call it trees on free leafset). This results from the fact that the traditional split-representation limits the possibility of extending those methods. This work has two major goals. The first one is to provide an alternative representation of common knowledge in phylogenetic trees as a representative set of subsplits. The second one is to provide an intuitive interpretation of a representative set, which would be convenient for researchers, specifically - for biologists. In order to complete the task, we adopt the frequent sets approach, well-known in the data-mining domain [3] for application in trees, but here we are considering issues specific for splits algebra. We define the frequent subsplit representation, and provide its interpretation as a profile of trees. The results prove that the proposed notions are reasonable.

---

* The research has been partially supported by grant No 3 T11C 002 29 received from Polish Ministry of Education and Science.

E. Marchiori and J.H. Moore (Eds.): EvoBIO 2008, LNCS 4973, pp. 95–105, 2008.

2   Common Knowledge Representation

2.1   Basic Notions

Leaf-labelled trees are very often represented as a set of splits [1] . Split (or Bipartition) $A|B$ (of a tree T with leafset L), corresponding to an edge e is a pair of leafsets A and B, which originated in splitting tree T into two disconnected trees whilst removing an edge e from a tree $T, A \cup B = L$. In this paper we will refer to the leafset of a given split s as $L(s)$.

For example tree $T1$ from Fig. 1, is built of the following splits:

$cd|abefghi, bcd|aefghi, abcd|efghi, hi|abcdefg, ghi|abcdef, fghi|abcde,$
$a|bcdefghi, b|acdefghi, c|abdefghi, d|abcefghi, e|abcdfghi, f|abcdeghi,$
$g|abcedfhi, h|abcdefgi, i|abcdefgh$

There have been many approaches aiming at constructing a tree that contains common knowledge of a particular group of trees. Among others: the maximum agreement subtree (MAST)[4], maximum compatible tree (MCT) [5] and a dozen of consensus trees [1,2]. The slightly different approach was using spectral analysis [6] to pick the best tree from input trees.

A strict consensus tree for example, is built out of splits that occur in all of the input trees. A Majority-rule consensus tree is built out of splits that occur in the majority of the input trees.

The proposed trees were not generally applicable but suitable for particular applications. For instance, consensus trees, the simplest and best known methods, were suitable only for trees with the same leafset. The other trees like MAST or MCT can theoretically be used for trees with a free leafset, however, they have other drawbacks. In a consensus tree one can choose, for example, a majority-rule consensus tree to avoid the situation in which one noisy tree will spoil the consensus result. In trees like MAST or MCT there is no such a simple solution. Additionally, MAST and MCT are in general NP-hard problems.

In [7] we have proposed the simplest possible approach to adapt consensus methods to trees with a free leafset, which, however, had the disadvantage of discarding some information apriori. In this paper we present a completely different and innovative approach to representing common knowledge in leaf-labelled trees. Our technique has all the advantages of consensus approaches: it allows choosing an arbitrary threshold of trees that are obliged to have the common knowledge; on the other hand, it is suitable also for trees with a free leafset and does not require discarding any data. In contrast to other methods known in literature, it does not necessarily result in one tree. The result of a method is a set of splits, which however, can easily be interpreted as a profile of trees, if the threshold is at least 50%.

2.2   Representative Splitset

Here we provide an alternative representation of common knowledge in trees.

Definition 1 (Restricted Split). *Split s_1 is a restricted version of split s_2 on the leafset z, if it is built with removing leaves not in z from s_2. $s_2^z = s_1$.*

Definition 2 (Restricted Split Equality(z-equality)). *Splits s_1 and s_2 are restrictedly equal on the leafset z, if those two splits after removing leaves not in z are equal.*

$$s_1 =^z s_2 \iff s_1^z = s_2^z. [?] \tag{1}$$

Definition 3 (Subsplit and supersplit). *Split s_1 is a subsplit of s_2, and s_2 is a supersplit of s_1, iff s_1 is restrictedly equal to s_2 on the leafset of s_1, and leafset of s_1 is a subset of the leafset of s_2.*

$$s_2 \subseteq s_1 \iff (s_1 =^z s_2) \wedge z = L(s_1) \wedge (L(s_1) \subseteq L(s_2)) \tag{2}$$

it can also be presented alternatively:

$$s_2(A|B) \subseteq s_1(C|D) \iff (A \subseteq D \wedge B \subseteq C) \vee (A \subseteq C \wedge B \subseteq D) \tag{3}$$

Definition 4 (Frequent subsplit). *Frequent subsplit s with support minsup in a profile of trees is a split that is a subsplit of at least one split in at least minsup of trees. The minsup parameter is called minimal support. It may be an absolute value which denotes the minimal number of trees, the split is supposed to be found in (as a subsplit). It can also be given as the relative value, given as a minimal percentage of tree, the split is supposed to be found in.*

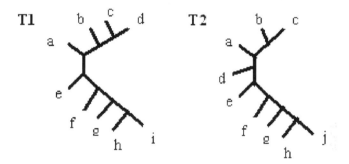

Fig. 1. Two leaf-labelled trees on different leafset

Consider the trees shown in the Fig. 1 , which are represented as follows:

$T1: cd|abefghi, bcd|aefghi, abcd|efghi, hi|abcdefg, ghi|abcdef, fghi|abcde,$
$a|bcdefghi, b|acdefghi, c|abdefghi, d|abcefghi, e|abcdfghi, f|abcdeghi,$
$g|abcedfhi, h|abcdefgi, i|abcdefgh$
$T2: bc|adefghj, abc|defghj, abcd|efghj, hj|abcdefg, ghj|abcdef, fghj|abcde,$
$a|bcdefghj, b|acdefghj, c|abdefghj, d|abcefghj, e|abcdfghj, f|abcdeghj,$
$g|abcedfhj, h|abcdefgj, j|abcdefgh$

In our approach we count the number of trees in which the split occurs (as a subsplit of any split), rather then counting the number of splits, of which it is a subsplit. For example, from Fig.1: $abcd|efgh$ has the support 2/2 (100%), because it occurs in both trees: in the first one as a subsplit of $abcd|efghi$, and in the second one as a subsplit of $abcd|efghj$. The argument for counting trees, rather than splits is that there might be some subsplits that occur frequently as subsplits of many splits, but only in one tree. Such trees are considered uninteresting.

Definition 5 (Representative splitset). *Representative splitset - a set that contains maximal frequent subsplits s, i.e. such that there is no other frequent subsplit s_2 that is also a supersplit of s.*

We may distinguish a strict representative splitset, that contains subsplits of all trees(minsup=100%), or a majority-rule representative splitset which contains subsplits of some splits in at least 50% of trees(minsup=50%).

For example from Fig.1 a strict splitset would be as follows:

$abcd|efgh, gh|abcdef, fgh|abcde, bc|aefgh, h|abcdefg,$
$a|bcdefgh, b|acdefgh, c|abdefgh, d|abcefgh, e|abcdfgh, f|abcdegh, g|abcedfh.$

The consensus methods for those trees would result in an empty splitset (because of a different leafset). The consensus tree built of z-restricted splits (see Def. 1) on a common leafset $(abcdefgh)$ will contain the three nontrivial splits $abcd|efgh, gh|abcdef, fgh|abcde$ plus the trivial split corresponding to each leaf from a common leafset.

For frequent splitset generation we do not use the classic approaches of generating all possible subsets of a given size, as it is done by means of frequent itemsets generation algorithms [3] .We generate the candidate subsplits from the input profile of trees, which is far more efficient, as the number of trees is relatively small (usually not more than a few hundreds). The generation procedure on Celeron M 1.6 Ghz with 512Mb of RAM memory for 100 trees with leafset of size 52 took 34 seconds, for 400 trees with leafset of size 52 - 136 seconds, for 100 trees with leafset of size 129 - 433 seconds. The procedure for the generation of frequent subsplits is as follows:

```
GENERATE_FS(TREE[] input, minsup)
  WHILE (C NOT EMPTY)
        C = whole, unique splits form input profile
        C=remove_infrequent_leaves(C)
        Divide C into frequent set  F
            and infrequent set IF according to minsup param
        F=REMOVE_NOT_MAXIMAL(F,RES)
        RES=sum(RES,F)
        C= GEN_CANDIDATES( IF );
  END WHILE

GEN_CANDIDATES( IF )
FOR EACH PAIR OF SPLITS(A|B,C|D) FROM "IF",
    THAT COME FROM DIFFERENT TREES
        X1=intersection(A,C), Y1=intersection(B,D),
        X2=intersection(A,D), Y2=intersection(B,C)
        CAND=sum(CAND, X1|Y1, X2|Y2)
  END FOR
return CAND
```

Where REMOVE_NOT_MAXIMAL(F,RES), removes splits from F, that have a supersplits in the so-far result RES.

3   Frequent Splitset Interpretation

Here we focus on the representative splitset interpretation. For example from section one, we had the following representative splitset: $abcd|efgh, gh|abcdef$, $fgh|abcde$, $bc|aefgh, h|abcdefg$, $a|bcdefgh, b|acdefgh$, $c|abdefgh$, $d|abcefgh$, $e|abcdfgh$, $f|abcdegh$, $g|abcedfh$. It is clear that from the given splits we cannot directly construct one tree because split $bc|aefg$ has a different leafset than other splits (d is missing). In order to provide an interpretation and visualization, below we list some properties of leaf-labelled trees.

Definition 6 (Split compatibility). *[8] Two splits $A_1|\overline{A_1}$ and $A_2|\overline{A_2}$ are compatible if*

1. *at least one of the intersections $A_1 \cap A_2$, $\overline{A_1} \cap \overline{A_2}$, $\overline{A_1} \cap A_2$ $A_1 \cap \overline{A_2}$ is empty*
2. *$\overline{A_i} = L - A_i$ and L is a set of all the possible leaves (the same for these two splits).*

Definition 7 (Compatible system). *A set of splits is compatible if every pair of splits is compatible [8]. Such a set is called a compatible system [9]. Every tree gives a compatible set of splits and every compatible set of splits gives a tree. (a known fact).*

Theorem 1. *If two splits s_1 and s_2 are compatible, then their subsplits meet the first condition of compatibility (their subsplits are semi-compatible).*

Proof. Let A_1 and A_2 be one side of the splits from s1 and s2, respectively. $A_1 \cap A_2 = \emptyset \Rightarrow \forall_{x,y} (A_1 \cap x) \cap (A_2 \cap y) = \emptyset$.

Theorem 2. *If two splits s_1 and s_2 are compatible then their z-restricted versions ($z \subseteq L(s_1) \cap L(s_2)$) are also compatible.*

Proof. The first condition of compatibility is met thanks to Theorem 1, z-restriction imposes the same leafset, and thus the second condition is also met.

Theorem 3. *Any two splits from a frequent splitset with a support greater than 50% are semi-compatible, i.e. one of the intersections $A_1 \cap A_2$, $\overline{A_1} \cap \overline{A_2}$, $\overline{A_1} \cap A_2$ $A_1 \cap \overline{A_2}$ is empty.*

Proof. If any two splits have a support of at least 50% then there must be a tree in the profile that contains supersplits (r_1, r_2) of both of them. $\forall_{r_1,r_2 \in S_R} \exists_{i,s_1 \in S_i, s_2 \in S_i} (r_1 \subseteq s_1 \wedge r_2 \subseteq s_2)$ And because r1 and r2 are compatible then s1 and s2 are also compatible (Theorem 1).

Theorem 4. *Any z-restricted subset of frequent splitset with the minimum support greater then 50% is a compatible set if $z \subseteq \bigcap L(s_i)$*

Proof. Emerges from Theorem 2 and 3.

From the above theorems we can derive the following conclusions:

Conclusion 1. *For each distinct leafset z from frequent splitset (FS) with a support greater then 50% a tree can be built. The tree is built on those splits from FS having a leafset as a superset of z. Therefore the frequent splitset (minsup>50%) can be represented as a set of trees. In particular, it affects the strict and majority-rule frequent set.*

Conclusion 2. *Each split from the frequent splitset discussed above will occur in at least one tree, in a restricted form.*

Conclusion 3. *Conclusions 1 and 2 are also true for a tree based on the intersection of all the distinct leafsets from frequent split-set.*

Conclusion 4. *The set of trees resulting from the frequent splitset will contain also a consensus tree, provided that the input dataset of trees were built on the same leafset.*

For a reconstruction of those trees, the procedure is as follows:

```
BUILD_TREES(FS)
FOR EACH distinct leafset z
    S=set of splits that are built on z
    X= set of z-restricted splits  that are built on  superset of L
    T=tree(sum(S,X))
END FOR
```

As, for example, the strict-frequent splitset of trees from Fig.1 contains splits built on two distinct leafsets: $abcdefg$, and $abcefg$, intersection of those leafsets is equal to the second leafset. Therefore, this strict-frequent splitset set will be illustrated by two trees.

Strict-frequent-set: $abcd|efgh, gh|abcdef, fgh|abcde, bc|aefgh, h|abcdefg,$ $a|bcdefgh, b|acdefgh, c|abdefgh, d|abcefgh, e|abcdfgh, f|abcdegh, g|abcedfh,$

Tree 1 $z = \{abcdefgh\}$:
Splits from Strict-frequent-set that are built on z (there are no splits built on superset): $abcd|efgh, gh|abcdef, fgh|abcde, h|abcdefg,$
$a|bcdefgh, b|acdefgh, c|abdefgh,$
$d|abcefgh, e|abcdfgh, f|abcdegh, g|abcedfh,$

Tree 2 $z = \{abcefgh\}$:
Splits from Strict-frequent-set that are built on z: $bc|aefgh,$
Splits from Strict-frequent-set that are built on superset of z:
$abcd|efgh, gh|abcdef, fgh|abcde, h|abcdefg, a|bcdefgh, b|acdefgh,$
$c|abdefgh, d|abcefgh, e|abcdfgh, f|abcdegh, g|abcedfh,$
And its z-restricted versions on $\{abcefgh\}$:
$abc|efgh, gh|abcef, fgh|abce, h|abcefg, a|bcefgh, b|acefgh, c|abefgh, e|abcfgh,$
$f|abcegh, g|abcefh,$

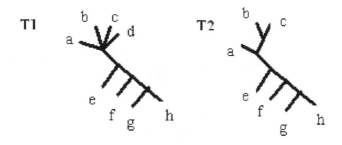

Fig. 2. Two trees built from strict frequent splitset of trees from Fig.2

Therefore T_2 is built of: $bc|aefgh, abc|efgh, gh|abcef, fgh|abce, h|abcefg,$
$a|bcefgh, b|acefgh, c|abefgh, e|abcfgh, f|abcegh, g|abcefh,$
For a more difficult example, let us look at trees T_1 and T_2 from Fig.3
T_1: $cd|abefgh, bcd|aefgh, abcd|efgh, gh|abcdef, fgh|abcde$
T_2: $bc|adefgh, abc|defgh, abce|dfgh, fg|abcdeh, fgd|abceh$
Plus the trivial splits in both trees: $a|bcdefgh, b|acdefgh, c|abdefgh,$
$d|abcefgh, e|abcdfgh, f|abcdegh, g|abcdefh, h|abcdefg$
The strict-frequent representative set: $a|bcdefgh, b|acdefgh, c|abdefgh,$
$d|abcefgh, e|abcdfgh, f|abcdegh, g|abcdefh, h|abcdefg, bc|aefgh, abc|efgh,$
$fgh|abce, fg|abcde$

Here we have three distinct leafsets: $\{abcdefgh\}$ $\{abcefgh\}$ $\{abcdefg\}$ and
the intersection: $\{abcefg\}$ Therefore as a visualization we present four trees on
these leafsets, as shown on Fig.4.

Such an approach yields a very good interpretation. Instead of choosing one
strict consensus tree, which often gives a star (like tree 1), or MAST - which
may significantly reduce the number of leaves and is inefficient for counting
(like tree 4), this method provides alternative interpretations based only on
the frequency of splits. Therefore it allows a full insight into the data without
using many different methods. For a large set of trees replacing 30 or more trees
with 4 different trees, representing a different perspective on the input data, is
far more informative than replacing it with, for example, one consensus tree.
Moreover, the consensus methods are useless for free-leafset data. The approach
is somehow similar to the Loose Consensus Tree described in literature [10] that
consists of splits that are compatible with at least one split in each tree. Such a
tree, however, had a disadvantage, since some splits from this tree may occur in
only one of the input trees. Our approach is better, as our splits, even when they
occur in complete form in only few trees still have to be direct subsplits of at
least a minsup of trees. For the sake of convenience, let's look at a majority-rule
frequent set(M-R-FS) for trees T_1, T_2, T_3 from Fig. 3.

T_1: $cd|abefgh, bcd|aefgh, abcd|efgh, gh|abcdef, fgh|abcde$
T_2: $bc|adefgh, abc|defgh, abce|dfgh, fg|abcdeh, fgd|abceh$
T_3: $ab|cdefgh, abc|defgh, abcd|efgh, gh|abcdef, fgh|abcde$

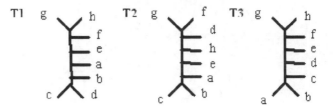

Fig. 3. Three leaf labelled trees on the same leafset

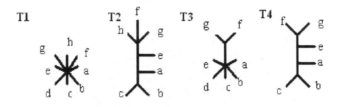

Fig. 4. Trees built from strict frequent splitset of trees T1 and T2 from Fig.3

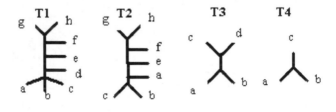

Fig. 5. Trees built from majority-rule frequent splitset of trees 2 from Fig.3

Plus trivial splits

M-R-FS=$abc|defgh$, $fgh|abcde$, $gh|abcdef$, $abcd|efgh$, $bc|aefgh$, $ab|dfgh$, $ab|cd$ plus trivial splits, which gives 3 leafsets $\{abcdefgh\}$ $\{abcefgh\}$ $\{abcd\}$, which gives us 3 trees.

4   Results

Here we present the results of counting strict and majority-rule frequent set. We have chosen the phylogenetic trees datasets: Camp(216 trees), Caesal(450 trees), pevccal(168 trees), kindly provided by Lee-San Wang, which contain leaf-labelled trees on the same leafset. The first experiment (shown in table 1 and table 2) presents information about the trees obtained from frequent splitset, as compared to consensus trees. The following information is provided: number of leaves in a tree / the informativity of the trees (i.e. the number of non-trivial splits) / how much the tree is resolved (i.e. the percentage of non-trivial splits in

Table 1. Trees on Strict Frequent Subsplit, for trees on the same leafset

k	Camp		Caesal		Pevccal	
	Strict FS	Strict CT	Strict FS	Strict CT	Strict FS	Strict CT
10	13/6/82% 11/6/89%	13/6/82%	51/44/95% 50/45/97%	51/44/95%	129/120/97% 127/121/98% 114/111/100% 7/3/90%	129/120/97%
20	13/6/82% 8/4/92%	13/6/82%	51/41/92% 50/42/94%	51/41/92%	129/118/96% 127/119/98% 114/109/99% 7/3/90%	129/118/96%
50	13/6/82% 8/4/92%	13/6/82%	51/39/90% 50/40/92% 40/31/92%	51/39/90%	129/116/96% 128/117/96% 128/117/96% 7/3/90%	129/116/96%
100	13/6/82%	13/6/82%	51/39/90% 50/34/92% 40/31/92%	51/39/90%	129/108/92% 128/109/93% 128/109/93% 119/104/94% 10/7/100% 7/3/90%	129/108/92%

Table 2. Trees on Majority-Rule Frequent Subsplit, for trees on the same leafset

k	Camp		Caesal		Pevccal	
	MR FS	MR CT	MR FS	MR CT	MR FS	MR CT
10	13/9/95%	13/9/95%	51/46/97%	51/46/97%	129/126/100%	129/126/100%
20	13/8/91%	13/8/91%	51/46/97% 50/46/98%	51/46/97%	129/126/100%	129/126/100%
50	13/7/86% 12/8/95% 12/7/90% 11/7/94% 10/6/94%	13/7/86%	51/46/97% 50/46/98% 49/45/98% 35/32/100%	51/46/97%	129/126/100%	129/126/100%
100	13/6/82% 11/6/89% 10/6/94% 9/5/93% 6/3/100%	13/6/82%	51/46/97%	51/46/97%	129/124/99% 7/4/100%	129/124/99%

given tree with respect to the number of non-trivial splits that the binary tree, built on this leafset would have). We provide these parameters, as the quality measure, because of the major drawback of consensus tree, which often produces star-tree (informativity 0, resolved in 0%) even though the trees are similar(see Fig. 4,tree T_1). Our technique tends not to have this disadvantage (other trees from Fig. 4). For the experiments we have used samples of different size (k). So, for example, for 10 trees from Camp dataset, table 1 contains the following result: 13/6/82% , 11/6/89% . This means that two different trees were constructed from

Table 3. Trees on Frequent Subsplit, for trees on free leafset, Camp dataset

k	10	20	50	100
Strict FS	12/5/80% 10/5/88%	12/5/80% 8/4/92%	12/5/80% 8/4/92%	12/5/80%
Strict CT	0/0/0	0/0/0	0/0/0	0/0/0
MR FS	12/8/95%	12/7/90%	12/6/85%, 11/7/94% 11/6/89%, 10/6/94% 10/6/94%	12/5/80%, 10/5/88% 9/5/93%, 9/5/93% 6/3/100%
MR CT	0/0/0	0/0/0	0/0/0	0/0/0

strict-frequent splitset. The first was built on 13 leaves, contained 6 non-trivial splits and was resolved in 82%, the other was built on 11 leaves, contained 6 non-trivial splits and was resolved in 89%. The results of the first experiment show that our method provides better common knowledge representation than consensus trees. The method used for the presented dataset produced from 1 to 6 trees from each sample, which gives a new and very interesting view on the input data. The single tree from the resulting trees was up to 12% more resolved than the consensus tree (sometimes fully resolved), and there was often more than one resulting tree. The results of the second experiment, which addresses the trees not built on the same leafset, are presented in table 3. For this experiment, we have used the same dataset as in the previous experiment, though slightly modified. For the strict frequent subsplit analysis, we have modified the label of one leaf in one tree. We have applied the label that does not occur in the input profile. For the majority-rule frequent subsplit analysis we have modified the same label of one leaf in all the input trees in such a way that it has become different in each tree(for example a is changed to a_1, a_2 ...). For lack of space, we provide only the results for Camp dataset. The results of the second experiment show that for very similar trees, albeit not built on the same leafset, the consensus methods completely fail, whilst our frequent subsplits representation provides very informative trees with almost the same quality as for trees with the same leafset.

5   Discussion

In this paper we have presented a new method of representing common knowledge in leaf-labelled trees as a set of frequent subsplits, together with its interpretation. Our technique has many advantages, which we have listed in this paper. The proposed solution has a very good interpretation, as it returns different, maximal sets of taxa that are connected with the same relations in the input trees. Such an approach often provides more interesting results than the tree containing whole taxa. The trees obtained from the frequent subsplit set contain at least one tree (which is a consensus tree if the trees are built on the same leafset). The rest of the resulting trees are more resolved than the trees obtained from the consensus methods. For trees that are not built on the same leafset, the method also provides very informative trees, whilst the consensus methods fail entirely. The algorithm for finding frequent leafset provides the result within reasonable time, which makes it useful in phylogenetic analysis. The proposed representation seems to be a very promising basis for the clustering of leaf-labelled trees, not necessarily built on the same leafset, which will be the part of our future work. The more efficient frequent subsplit generation methods will also be addressed.

References

1. Bryant, D.: Building trees, hunting for trees, and comparing trees. Theory And Method. In: Phylogenetic Analysis. Ph.D Thesis University of Canterbury (1997)
2. Bryant, D.: A classification of consensus methods for phylogenetics. In: Bioconsensus. DIMACS Series in Discrete Mathematics and Theoretical Computer Science, vol. 61, pp. 163–184. AMS Press, New York (2002)
3. Agrawal, R., Srikant, R.: Fast algorithms for mining association rules. In: Proc. of the 20th Int'l. Conference on Very Large Databases (VLDB 1994), Santiago, Chile, pp. 487–499 (1994)
4. Finden, C.R., Gordon, A.D.: Obtaining common pruned trees. Journal of Classification 2, 255–276 (1985)
5. Ganeshkumar, G., Warnow, T.: Finding a Maximum Compatible Tree for a Bounded Number of Trees with Bounded Degree is Solvable in Polynomial Time. In: Gascuel, O., Moret, B.M.E. (eds.) WABI 2001. LNCS, vol. 2149, pp. 156–163. Springer, Heidelberg (2001)
6. Hendy, M.D., Penny, D.: Spectral analysis of phylogenetic data. Journal of Classification 10, 5–24 (1993)
7. Koperwas, J., Walczak, K.: Clustering of leaf labeled-trees on free leafset. In: Kryszkiewicz, M., Peters, J.F., Rybinski, H., Skowron, A. (eds.) RSEISP 2007. LNCS (LNAI), vol. 4585, pp. 736–745. Springer, Heidelberg (2007)
8. Tubingen, U., Huson, D.: Algorithms in Bioinformatics, Lecture Notes, http://www-ab.informatik.uni-tuebingen.de
9. Holland, B., Moulton, V.: Consensus networks: A method for visualising incompatibilities in a collection of trees. In: Benson, G., Page, R.D.M. (eds.) WABI 2003. LNCS (LNBI), vol. 2812, pp. 165–176. Springer, Heidelberg (2003)
10. Stockham, C., Wang, L.S., Warnow, T.: Statistically based postprocessing of phylogenetic analysis by clustering. Bionformatics 18, 285–293 (2002)
11. Koperwas, J., Walczak, K.: Clustering of leaf-labelled trees. In: Beliczynski, B., Dzielinski, A., Iwanowski, M., Ribeiro, B. (eds.) ICANNGA 2007. LNCS, vol. 4431, pp. 702–710. Springer, Heidelberg (2007)
12. Bryant, D.: The Splits in the Neighborhood of a Tree. Ann. Combinatorics 8 (2004)
13. Amenta, N., Klingner, J.: Case study: Visualizing sets of evolutionary trees. In: 8th IEEE Symposium on Information Visualization, pp. 71–74 (2002)

Inference on Missing Values in Genetic Networks Using High-Throughput Data

Zdena Koukolíková-Nicola[1], Pietro Liò[2], and Franco Bagnoli[3]

[1] Fachhochschule Nordwestschweiz, Hochschule für Technik, Steinackerstrasse 5,
CH-5210 Windisch, Switzerland
[2] The Computer Laboratory, University of Cambridge, William Gates Building,
15 JJ Thomson Avenue, Cambridge CB3 0FD, UK
[3] Department of Energy, University of Florence,
Via S. Marta, 3 50139 Firenze.
Also CSDC and INFN, sez. Firenze

Abstract. High-throughput techniques investigating for example protein-protein or protein-ligand interactions produce vast quantity of data, which can conveniently be represented in form of matrices and can as a whole be regarded as knowledge networks. Such large networks can inherently contain more information on the system under study than is explicit from the data itself. Two different algorithms have previously been developed for economical and social problems to extract such hidden information. Based on three different examples from the field of proteomics and genetic networks, we demonstrate the great potential of applying these algorithms to a variety of biological problems.

1   Introduction

Current high-throughput techniques produce large amounts of biological data that can usually be represented in the form of large matrices. For example, experimental results from DNA or protein (protein-protein, protein-ligand or enzyme-substrate) microarray assays are assembled in a matrix, where the genes/proteins/ligands constitute the rows and the different experimental conditions under which they were probed the columns, see for example [10].

Two papers have recently proposed algorithms to extract meaningful relationships from matrix data describing large networks [7,1]. [7] have demonstrated that it is for example possible for an advisor service to recommend new books to their customers, based on their incomplete knowledge about the overlap of customers' preferences. However, in many practical cases, the data needed by the previous algorithm are unaccessible. For instance, it is very hard to obtain data about preferences, while it is possible to know customers' opinions about already read books. [1] have shown that it should be in principle possible to extract information about quantities hidden in such opinion networks. Suppose we have data on a large network of proteins either interacting with each other or binding to many different ligands. Such information can in principle be interpreted as data on different proteins' "opinion" on other proteins or different

E. Marchiori and J.H. Moore (Eds.): EvoBIO 2008, LNCS 4973, pp. 106–116, 2008.

proteins' "opinion" on different ligands (or *vice versa*). Thus it should theoretically be possible to anticipate or "predict" unknown interactions, which were not yet probed for, and which could then be verified experimentally. Moreover, this method should also be useful either for the interpolation of missing values or the estimation of the consistency of the entries in a data matrix with some model hypothesis. Since many analysis methods, such as principal components analysis or singular value decomposition, require complete matrices, different imputation methods were developed and tested by [10] for use with DNA microarray data.

In this work, we first introduce the two knowledge networks algorithms we used to tackle important problems in the field of proteomics. We then demonstrate the power and potential of this approach on three specific examples on which those algorithms have successfully been applied: the prediction of amino acid contacts of a protein structure, inference of substrate-enzyme relationships for a novel enzyme given a set of enzymes and their substrates and inference on cellular interactions mediated by cytokines based on a network, manually compiled from literature [3].

2 System and Methods

2.1 Maslov-Zhang's (MZ) Algorithm

Demonstration. We first checked the prediction potential of the MZ-algorithm [7] devising three very simple networks of 6 circularly or partly connected nodes (Fig. 1A.). Apart from the explicitly defined direct connections between immediately adjacent nodes, depending on their network architecture, these example networks also contain several implicit indirect connections between non-adjacent nodes. They actually only differ in the way node D can be reached from the other nodes: Whereas in the first network *I* node D is separated from either node C or node E by only a distance of one step, node D can only be indirectly reached from node C in the second network *II*, passing by nodes B, A, F, and E, such that the two nodes are separated by a distance of five steps. On the contrary, the architecture of the third network *III* is such that node D lies isolated and can't be reached from any of the other nodes, which is defined by a distance of zero steps.

These three networks were each represented by a symmetrical connection matrix, where directly connected nodes are assigned a connection value of one and all other potential indirect connections obtain the initial value of zero. MZ-algorithm was then applied to each matrix as following: at every round, we kept the values of the known direct connections, but updated all the values of the "unknown" indirect links. To probe the power of the algorithm, we used different numbers of positive eigenvalues (1 to 4), thus exploiting different amounts of "knowledge" about the networks. As an example, results for the case, where only the two largest eigenvalues were used, are shown for the network *II* and *III* (Fig. 1B. and C). In all cases, *i.e.* all three networks and all numbers of eigenvalues tested (except for network *III*, when using the largest

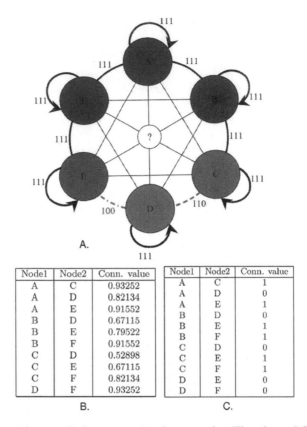

A.

Node1	Node2	Conn. value
A	C	0.93252
A	D	0.82134
A	E	0.91552
B	D	0.67115
B	E	0.79522
B	F	0.91552
C	D	0.52898
C	E	0.67115
C	F	0.82134
D	F	0.93252

B.

Node1	Node2	Conn. value
A	C	1
A	D	0
A	E	1
B	D	0
B	E	1
B	F	1
C	D	0
C	E	1
C	F	1
D	E	0
D	F	0

C.

Fig. 1. MZ-algorithm applied to very simple networks. The three different networks tested (I,II,III) are schematically shown in a single drawing (A.). All direct links between adjacent nodes are initially assigned connection values: *e.g.* 110 means that network *I* and *II* have a value of 1 and network *III* a value of 0 associated with that link. The tables correspond to networks *II* (B.) and *III* (C.). Columns 1 and 2 define the two nodes whose connection was predicted; column 3 contains estimated connection values; number of positive eigenvalues used = 2.

eigenvalue only), the algorithm correctly detected any indirect connections between non-adjacent nodes via other nodes and assigned to them values, which slightly decrease with the increasing distance. This effect is more pronounced for networks *II* and *III*, as not all indirect connections are possible in those cases. Most important, no links are detected between those nodes that can't be connected by any indirect way (compare tables B. and C. in Fig. 1).

Application to Contact Maps. The contact maps were constructed using Shanahan's program (http://www.biochem.ucl.ac.uk/~shanahan) that was run at several maximal distance thresholds between 2 and 8 Å. The lowest threshold, at which an interaction between two amino acids was found, was used as the approximate distance of that contact in Å. We then assigned values to the different amino acid interactions applying the following rule: value = 8 - (distance in Å - 2). This somewhat arbitrary rule attributes higher weight to interactions at short range, allowing a finer "rating" of the interactions. The following algorithm adapted from [7] was then applied to each position of the contact matrix representing a specific protein domain: (1) To simulate unknown values, construct the initial CM matrix by deliberately setting the values of one specific position and its corresponding symmetric position to zero. (2) Normalise the thus slightly changed CM matrix per row, giving matrix N_CM. (3) Diagonalise matrix N_CM, and construct the N_CM' by keeping only the M largest (positive) eigenvalues and eigenvectors of the old matrix, while setting the remaining eigenvalues to zero. (4) Construct a new refined approximate N_CM matrix by copying the unknown elements from N_CM', while resetting the rest to their exactly known values. (5) Go to the step (3). Repeat until either the values being predicted do not change considerably or a limit number of rounds of the algorithm is reached. As a rule, the number M of positive eigenvalues was varied. M_{eff}, the effective number of eigenvalues was estimated using the formula $M_{\text{eff}} \approx K/N$, where K is the number of known contacts and N is the size of the CM matrix, i.e. the total number of amino acids in the protein domain under investigation [7].

Application to Cytokine Networks. We translated the cytokine network connections, described in Fig. 4 of [3], into a square data matrix assigning to all mutual connections a value of 2 and to all one-way connections a value of 1. MZ-algorithm was applied as described in the previous section with a slight modification of step (3): Decompose the connection matrix N_CM into its singular values and vectors, and construct the N_CM' by keeping only the M largest singular values of the old matrix. All other steps remain unchanged. The number M of singular values used was varied from $1 - 29$.

2.2 De Gustibus Algorithm

In each calculation, a specific enzyme-ligand entry in the matrix containing the binding energies calculated by [6] was set to zero and the thus resulting new matrix was normalized as follows: The matrix average was subtracted from all values, which were then made positive by subtracting from each of them the largest negative value found in the whole matrix. Finally, all matrix entries were divided by the largest one of all. Each time, a correlation matrix was then calculated applying the equation 1 from [1]:

$$C_{i,j} = \frac{\sum_{n=1}^{N}(s_{i,n} - \overline{s_i})\,(s_{j,n} - \overline{s_j})}{\sqrt{\sum_{n=1}^{N}(s_{i,n} - \overline{s_i})^2 \; \sum_{n=1}^{N}(s_{j,n} - \overline{s_j})^2}} \tag{1}$$

where $C_{i,j}$ denotes the Pearson correlation coefficient between enzyme i and j; N is the number of different ligands in the interaction matrix; $s_{i,n}$ and $s_{j,n}$ correspond to the binding energies of ligand n to the enzymes i and j, respectively; and $\overline{s_i}$ and $\overline{s_j}$ denote the binding energies for either enzyme i or j, averaged over all ligands. We did not include on purpose any binding energy values involving the ligand in question into the correlation calculations, in order not to skew the coefficients by the "missing" value. In our original attempt, the binding energy of the given enzyme-ligand interaction was "predicted" using the equation 3 from [1]:

$$\tilde{s}^*_{m,n} = \frac{1}{M} \sum_{i=1}^{M} C_{m,i} \, s_{i,n} \tag{2}$$

where $\tilde{s}^*_{m,n}$ denotes the estimated value of binding energy of ligand n to enzyme m; M is the total number of different enzymes in the interaction matrix; $C_{m,i}$ stands for correlation coefficient between enzyme m and enzyme i; and $s_{i,n}$ represents binding energy of ligand n to enzyme i. We omitted the zero value from the calculation, again not to influence it by the "non-existing" value. This was repeated for each enzyme-ligand pair of the matrix. In our final approach, the relative weight w_k for each of the K enzymes contributing to the estimation of a value for enzyme A is calculated as follows. An ordered list of all the other enzymes, which are positively correlated to enzyme A, is assembled according to their correlation coefficients. w_k is then calculated for the K enzymes with the highest coefficients:

$$w_k = \frac{C_{k,i}}{\sum_{k=1}^{K} C_{k,i}} \tag{3}$$

where k designates the index of the first K enzymes in the ordered list; i is the index of enzyme A; and $C_{k,i}$ denotes the correlation coefficient of the kth enzyme to the enzyme i. A "normalised value" is calculated using the following formula:

$$\tilde{s}'_{m,n} = \sum_{k=1}^{K} w_k \, s'_{k,n} \tag{4}$$

where $\tilde{s}'_{m,n}$ is the estimated "normalised value" of the binding energy of enzyme m to ligand n; and $s'_{k,n}$ denotes the "normalised value" of the interaction between ligand n and the kth enzyme in the ordered list. The thus "predicted" values were then re-transformed to the final binding energy according to the normalising factors used and then compared to the original ones.

Statistical Assessment. To assess statistical relevance of the results produced by the *De Gustibus* method, we performed 1000 randomisation-predic-tion experiments and compared them to the original results: At each round, the matrix containing the enzyme-substrate binding energies was randomised anew, *i.e.* the values of the matrix were randomly redistributed. The "predictions" were then performed exactly as described above. The results of each such experiment were grouped into different error categories depending on the extent of the absolute difference between the "estimated" and the original value for the respective enzyme-substrate pair. For each category, the mean value and the standard deviation was calculated from all the randomisation experiments. Those can be compared to the values of the corresponding categories compiled from the original estimations.

3   Results and Discussion

3.1   Contact Maps of Compact Proteins

Encouraged by the results of our prior tests of the MZ-method on simple artificial networks (see System and Methods), we then tackled our first example problem. The 3D-structure of a protein can be represented as a contact map (CM). In case of globular proteins, Porto and collaborators have shown that the principal eigenvector of a contact map is equivalent to the CM itself, and therefore basically contains all the information about the 3D-structure of a protein. Thus it seems reasonable to view contact maps as 3D-structure knowledge networks and use their eigenvalues to reconstruct unknown or uncertain entries of the contact matrix. Moreover, CMs are symmetric square matrices by definition. For these reasons, MZ-algorithm is ideal to "predict" or validate the internal contacts between different amino acids of a globular polypeptide chain, based on the known contacts.

We used two well studied proteins: the globular human protein haemoglobin (PDB ID: 2HHB) [2] and the protein HisA from *Thermotoga maritima* (PDB ID: 1QO2), which was shown to be folded as a compact β/α barrel [4]. We constructed the CMs with the help of Shanahan's program (http://www.biochem. ucl.ac.uk/~shanahan), assigning values to the different contacts depending on the estimated distance between the individual amino acids: the shorter the distance in Å, the higher the value assigned. Shanahan's program was ideal for our purpose since it exclusively uses the data in the PDB file without any further interpretation or any evaluation of the 3D-structure of the protein. Taking one position after the other of the CM, we tested whether all the known contacts and their respective values could be reconstructed using the information contained within the rest of the CM. As was theoretically shown by [7], the density of the known interactions in the CM must be higher than a certain threshold p_2, so that all the unknown contacts are completely determined by the information contained in the known ones. Below p_2, the algorithm performs rather poorly. However, this threshold can't be estimated for real problems, since the number of relevant components determining the actual contacts is unknown. For the same reason, the number of positive eigenvalues required for the calculations thus cannot be determined precisely. That's why, as a first approximation, we employed Maslov and Zhang's conjecture to estimate the effective number of eigenvalues M_{eff} to be used. As the CM for the whole protein is usually quite sparse, only certain "domains" of the CM matrix could be used for predictions: as a rule of thumb, domains with $\geq 50\%$ of known contacts were analysed. For obvious reasons, we used symmetrical domain-CMs, *i.e.* the amino acids involved were the same in both directions of the matrix. Usually domains of 20–25 amino acids were suitable for analysis. For each position (i.e. each putative contact between two amino acids) of such a domain-CM, the contact value was "predicted" by MZ-algorithm, keeping about M_{eff} of the positive eigenvalues (see System and Methods). This procedure was repeated for all positions in the domain-CM and the so estimated values were compared to the original ones (see Table 1). We

Table 1. Contact predictions for two different protein domains (Remark: Values less than 2 indicate a distance between the amino acids greater than 8Å; values greater than 2 indicate a distance less than 8Å)

Protein HisA (PDB ID: 1QO2) Domain aa53 - aa73		Estimated $M_{eff} \approx 5.6$
For cases where original value $\neq 0$	$M_{eff} = 5$	$M_{eff} = 6$
Percent of cells with predicted value $\leq 10\%$ of the original one	36.3	37.6
Percent of cells with predicted value $\leq 20\%$ of the original one	67.5	62.9
Percent of cells with predicted value $\leq 50\%$ of the original one	91.1	91.1
Total number of cells	237	237
For cases where original value $= 0$	$M_{eff} = 5$	$M_{eff} = 6$
Percent of cells with predicted value < 1	80.4	84.3
Percent of cells with predicted value > 1 AND < 2	16.7	11.8
Percent of cells predicted with value > 2	2.9	3.9
Total number of cells	204	204
Protein Haemoglobin (PDB ID: 2HHB) Domain aa15 - aa40		Estimated $M_{eff} \approx 6.5$
For cases where original value $\neq 0$	$M_{eff} = 6$	$M_{eff} = 7$
Percent of cells with predicted value $\leq 10\%$ of the original one	32.8	35.8
Percent of cells with predicted value $\leq 20\%$ of the original one	60.7	66.6
Percent of cells with predicted value $\leq 50\%$ of the original one	95.9	95.9
Total number of cells	338	338
For cases where original value $= 0$	$M_{eff} = 6$	$M_{eff} = 7$
Percent of cells with predicted value < 1	81.7	82.8
Percent of cells with predicted value > 1 AND < 2	16.6	15.4
Percent of cells predicted with value > 2	1.8	1.8
Total number of cells	338	338

demonstrated for several such domains of the proteins, haemoglobin and HisA, that 80–90% of the "predicted" values deviate by less than 50% from the original value. Summary of the results for two of the different protein domains are shown in Table 1.

3.2   Cytokine Networks

As a further example of application, we used the cytokine network from [3] which consists of cells as nodes (29 in total) and cytokine connections as edges (418 in total), disregarding as yet which particular cytokines are actually involved. We assigned to mutual and one-way connections different values (2 and 1, respectively). For each position of the matrix we tested whether all the known connections and their respective values could be reconstructed. Since the data matrix representing the current knowledge is not completely symmetrical, we had to modify slightly the "prediction" algorithm: we decomposed each matrix

Table 2. Predictions for cellular interactions mediated by cytokines

	nb of singular values $= 1$
Percent of cells with predicted value $\leq 5\%$ of the original one	9.1
Percent of cells with predicted value $\leq 10\%$ of the original one	17.5
Percent of cells with predicted value $\leq 20\%$ of the original one	33.5
Percent of cells with predicted value $\leq 50\%$ of the original one	78.7
Percent of cells with predicted value $\leq 100\%$ of the original one	100.0
Total number of cells	418

into its singular values and corresponding vectors, as by definition this always results in real positive singular values. We tested different numbers of singular values for the matrix reconstructions $(1 - 29)$, but interestingly only one or two singular values gave reasonable results. As the number of singular values was increased, the results quickly deteriorated (data not shown), probably due to the inclusion of the noise contained in the vectors corresponding to small singular values. Assuming that the mutual and one-way connections in the matrix are reliable, whereas absence of any connections may at present be due to the fact that for whatever reason it has yet not been discovered, we based our reliability estimation of the results exclusively on the known connections. Table 2 shows that using one singular value about 80% of the "predicted" connection values deviate by less than 50% from their original value.

3.3 Correlation Matrices

Bagnoli and collaborators extended the MZ-approach to knowledge networks where direct interactions between different nodes are unknown, but can be deduced by constructing a similarity correlation matrix between them. We applied their algorithm to the data generated by [6], where they studied the selectivity of recognition between *E. coli* enzymes and their cognate substrates using standard docking calculations. The data set we chose comprises binding energies calculated for 27 enzymes to 119 different ligands (including the 27 cognate substrates of the enzymes). It actually constitutes a subset of the raw results described by [6]. Our data set is particularly suitable for demonstrating the prediction or validation potential of the correlation matrices, since it represents a completely filled matrix. The data may be regarded as "opinion of the 27 different enzymes on the 119 different ligands", and a correlation matrix between the 27 enzymes' opinions can easily be constructed. To test the predictive power of the *De Gustibus* method, we calculated one by one the binding energy for every enzyme-ligand pair, based on the other binding energies (except the one for the pair in question) and on the corresponding correlation matrix between all the enzymes of the data set. This approach allowed us to address the question, whether the data set, which originates from pure calculations indeed, shows internal consistency. This was particularly interesting since surprisingly, the docking results [6] indicated that the cognate enzyme-ligand complex does rarely have the lowest binding energy from all the complexes involving manifold alternative substrates. Our first approach, applying the Eq. (2) as such, did not yield very good results, since about 36% of the "predicted" values differed by more than 30% from the originally calculated binding energy (data not shown). We reasoned that this might be due to the fact that in that equation the prediction for the ligand in question is based on the values for all the other enzymes, regardless of whether those enzymes are tightly correlated or not to the one in question. We thus adopted a strategy similar to the one employed by [10] to estimate missing values in DNA microarray studies. Suppose an enzyme A that has one missing value for the interaction with ligand 1. Based on the correlation coefficients, a weighted average of the values involving enzymes most similar to enzyme A

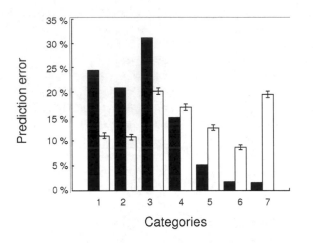

Fig. 2. Accuracy of "predicted" enzymes-ligands binding energies. Comparison of the results from the original predictions (black bars) and those compiled from the 1000 randomisation experiments (white bars). In the former case, the 3213 predictions were simply grouped into different error categories, whereas in the latter, the mean values and corresponding standard deviations were calculated for the different categories and are shown (white bars and respective black bars). Percentage of estimations belonging to the different categories (Y-axis). Categories (X-axis): absolute difference between the estimated and original value comprises $< 5\%$ of the original value (1), difference is $\geq 5\%$ but $< 10\%$ (2), difference is $\geq 10\%$ but $< 20\%$ (3), difference is $\geq 20\%$ but $< 30\%$ (4), difference is $\geq 30\%$ but $< 40\%$ (5), difference is $\geq 40\%$ but $< 50\%$ (6), and difference is $\geq 50\%$ (7).

is used for "prediction". Actually, only the K enzymes, exhibiting the highest positive correlation coefficient to enzyme A, are included in the estimation. A relative weight w_k is designated to each of those enzymes. The contribution of each of the K enzyme-ligand interaction data is thus weighted by the value of the correlation coefficient between the enzyme A and the enzyme in question. Results of our final approach using $K = 10$ are shown in Fig. 2. In this special case, where we deal with a full "opinion" matrix of 3213 interactions and are estimating one of them at a time, about 91% of the individual "predicted" values did not deviate by more than 30% (and about 98% did not differ by more than 50%) from the originally calculated binding energy. Compare these results to those compiled from 1000 randomisation experiments, described in System and Methods (Fig. 2).

3.4   Comparison and Work in Progress

There is a tremendous lack of algorithms for inferring missing values in data produced by high-throughput techniques. Moreover, some types of data may require specialised algorithms. We compared our general algorithm with the KNNimpute program of [10] that has been developed and tested for imputation of missing values in gene microarray data. Applying the *De Gustibus* and the KNNimpute method on the same data sets (from [9]) as [10], we obtained very similar results (*e.g.* 95–96% of the imputed values differed by less than 50% from the original values). The availability of huge amount of data in form of replicated data sets allows to simultaneously approach the problems of dimensionality of biological data, *i.e.* the number of variables may in some cases exceed the number of responses by a factor of tens, and the problem of their dependencies (see also [5]). We aim at implementing statistical techniques such as Markov chain Monte Carlo and Bayesian inference to identify which values from the available data should be used for optimal inference on the missing ones. These methods, which are beyond the scope of this paper, are just beginning to have a significant impact in high-throughput data statistics.

4   Conclusions

We have investigated the possibilities and the potential of applying the MZ and *De Gustibus* algorithm to several types of problems which arise in proteomics. We have demonstrated that indeed both algorithms are useful. The MZ-algorithm was successfully applied not only on our toy example but also on the problem of validating uncertain amino acid contacts of a protein tertiary structure. We further showed that, in principle, it is possible to suggest yet unknown connections in a network, such as the cytokine one, based on the present knowledge contained within. The *De Gustibus* algorithm was tested on results from *in silico* docking experiments of several *E. coli* enzymes to a multitude of different ligands. Surprisingly, the enzymes appeared to be quite promiscuous since many ligands "bound" at least as well if not better than their cognate substrates. Since docking calculations have their limitations and imperfections, it was particularly interesting to assess the reliability of those results. Indeed, our study strengthens the conclusions by [6] in that it demonstrates that their docking results constitute a set of binding energy values overall internally consistent with each other. The limitations of the two algorithms lie clearly in the fact that the matrices may not be too sparse in order to allow reliable predictions. However, with the ever growing amount of biological data, we are confident that they will be useful for an increasing number of applications.

References

1. Bagnoli, F., Berrones, A., Franci, F.: De gustibus disputandum (forecasting opinions by knowledge networks). Physica A 332, 509–518 (2004)
2. Fermi, G., Perutz, M.F., Shaanan, B., Fourme, R.: The crystal structure of human deoxy haemoglobin at 1.74 Å resolution. J. Mol. Biol. 175(2), 159–174 (1984)
3. Frankenstein, Z., Alon, U., Cohen, I.: The immune-body cytokine network defines a social architecture of cell interactions. Biology Direct 1(32), 1–15 (2006)
4. Lang, D., Thoma, R., Henn-Sax, M., Sterner, R., Wilmanns, M.: Structural Evidence for Evolution of the α/β Barrel Scaffold by Gene Duplication and Fusion. Science 289, 1546–1550 (2000)

5. Liò: Dimensionality and dependence problems in statistical genomics. Brief Bioinform 4, 168–177 (2003)
6. Macchiarulo, A., Nobeli, I., Thornton, J.: Ligand selectivity and competition between enzymes in silico. Nature Biotechnology 22(8), 1039–1045 (2004)
7. Maslov, S., Zhang, Y.-C.: Extracting Hidden Information from Knowledge Networks. Physical Review Letters 87(24), 248701_1–248701_4 (2001)
8. Porto, M., Bastolla, U., Roman, H.E., Vendruscolo, M.: Reconstruction of Protein Structures from a Vectorial Representation. Physical Review Letters 92(21), 218101_1–218101_4 (2004)
9. Spellman, P., Sherlock, G., Zhang, M., Iyer, V., Anders, K., Eisen, M., Brown, P., Botstein, D., Futcher, B.: Comprehensive Identification of Cell Cycle-regulated Genes of the Yeast *Saccharomyces cerevisiae* by Microarray Hybridization. Molecular Biology of the Cell 9, 3273–3297 (1998)
10. Troyanskaya, O., Cantor, M., Sherlock, G., Brown, P., Hastie, T., Tibshirani, R., Botstein, D., Altman, R.: Missing value estimation methods for DNA microarrays. Bioinformatics 17(6), 520–525 (2001)

Mining Gene Expression Patterns for the Discovery of Overlapping Clusters

Patrick C.H. Ma[*] and Keith C.C. Chan

Department of Computing, The Hong Kong Polytechnic University,
Hung Hom, Kowloon, Hong Kong, China
cschma@comp.polyu.edu.hk

Abstract. Many clustering algorithms have been used to identify co-expressed genes in gene expression data. Since proteins typically interact with different groups of proteins in order to serve different biological roles, when responding to different external stimulants, the genes that produce these proteins are expected to co-express with more than one group of genes and therefore belong to more than one cluster. This poses a challenge to existing clustering algorithms as there is a need for overlapping clusters to be discovered in a noisy environment. For this reason, we propose an effective clustering approach, which consists of an initial clustering phase and a second re-clustering phase, in this paper. The proposed approach has several desirable features as follows. It makes use of both local and global information inherent in gene expression data to discover overlapping clusters by computing both a local pairwise similarity measure between gene expression profiles and a global probabilistic measure of interestingness of hidden patterns. When performing re-clustering, the proposed approach is able to distinguish between relevant and irrelevant expression data. In addition, it is able to make explicit the patterns discovered in each cluster for easy interpretation. For performance evaluation, the proposed approach has been tested with both simulated and real expression data sets. Experimental results show that it is able to effectively uncover interesting patterns in noisy gene expression data so that, based on these patterns, overlapping clusters can be discovered and also the expression levels at which each cluster of genes co-expresses under different conditions can be better understood.

Keywords: Data Mining, clustering, bioinformatics, gene expression data analysis.

1 Introduction

Given a database of records each characterized by a set of attributes, the clustering problem is concerned with the discovery of interesting record groupings based on attribute values [1]. To deal with the increasing amount of gene expression data produced by the DNA microarray technology [2], [3] existing clustering algorithms [4], [5], [6] have been used to the identification of co-expressed genes [7], [8], [9]. Since

---

[*] Corresponding author.

E. Marchiori and J.H. Moore (Eds.): EvoBIO 2008, LNCS 4973, pp. 117–128, 2008.
© Springer-Verlag Berlin Heidelberg 2008

co-expressed genes have similar transcriptional response to external stress, they exhibit similar expression patterns and could have similar or related biological functions. Effective clustering of gene expression data can therefore allow these patterns to be more easily discovered.

While many clustering algorithms have been used successfully with gene expression data, it should be noted that most of them (e.g., the k-means algorithms) usually perform their tasks under the assumption that each gene belongs only to one cluster. Such an assumption can sometimes be an over-simplification of the great biological complexity underlying the gene expression process. As many proteins have multiple functional roles in a cell, they have to interact with different groups of proteins to fulfill them. The genes that produce these proteins are therefore expected to co-express with different groups of genes in order to meet the varying demands of a cell. In other words, depending on which experimental conditions being investigated, each gene may have similar expression patterns with different groups of genes in other clusters and they can, therefore, belong to more than one cluster. This poses a challenge to existing clustering algorithms as they need to tackle two difficult problems: (i) they need to handle overlapping clusters, and (ii) they need to discover overlapping clusters in the presence of various forms of data inaccuracies and variations arising not only from genetic variations and impurity of tissue samples but also from such processes as the production of the DNA array, the preparation of the samples, the hybridization of experiments, and the extraction of the hybridization results [10], etc. In light of these problems, some recent attempts have been made to use the fuzzy k-means algorithms [11], [12] and the biclustering algorithms [13], [14] to discover overlapping clusters in gene expression data with limited success. In addition, many existing clustering algorithms do not make explicit the patterns discovered in a data set during the clustering process. To better understand and interpret the clustering results, a separate technique is usually required for patterns underlying each discovered cluster to be uncovered explicitly.

To discover overlapping clusters in noisy gene expression data effectively, we propose a clustering approach, which consists of an initial clustering phase and a second re-clustering phase, in this paper. The proposed approach makes use of both local and global information hidden in gene expression data when deciding which genes to be grouped together. In the first phase, local information is extracted by computing a pairwise similarity measure between gene expression profiles so as to detect for underlying clustering structures (the initial clusters). In the second phase, global information is obtained through discovering interesting patterns in the initial clusters. These patterns are identified by differentiating between expression data that are relevant for the clustering process from those that are not relevant. In doing so, a probabilistic interestingness measure is used. If an expression level is relevant in determining whether or not a gene should belong to a particular cluster, it is reflected by the interestingness measure. Since this measure is probabilistic, it can work effectively even when the data being dealt with contains incomplete, missing, or erroneous values. Once the relevant expression data are identified, the discovered patterns can be made explicit for easy interpretation. Based on the discovered patterns, the cluster memberships of genes in each initial cluster are re-evaluated to determine if they should remain in the same cluster or be assigned to more than one. With the proposed approach, overlapping clusters can then be effectively discovered in noisy gene expression data.

The rest of this paper is organized as follows. In Section 2, the proposed clustering approach is described in details. The effectiveness of this approach has been evaluated through the experiments with both simulated and real expression data sets. The experimental set-up, together with the results, is discussed in Section 3. In Section 4, we give a summary of the paper.

2 The Proposed Clustering Approach

To describe the proposed clustering approach, let us assume that we are given a set of gene expression data, G, consisting of the data collected from N genes in M ments carried out under different sets of experimental conditions. Let us represent the data set as a set of N genes, $\mathbf{G} = \{g_1, \ldots, g_i, \ldots, g_N\}$, with each gene, g_i, $i =$ characterized by M attributes, $E_1, \ldots, E_j, \ldots, E_M$ whose values, $e_{i1}, \ldots, e_{ij}$, where $e_{ij} \in \text{domain}(E_j)$ represents the expression value of the ith gene under the jth experimental condition.

Phase 1: An Initial Clustering Phase
To find the initial clusters, any traditional (crisp) clustering algorithms such as the means algorithm can be used. As mentioned before, these algorithms have been used successfully with gene expression data, we therefore make use of the strengths of existing algorithms in this phase. In addition, researchers can easily extend their clustering algorithms to discover overlapping clusters by only integrating the proposed re-clustering technique as presented in Phase 2 below. Here, we used the popular k-means clustering algorithm [5] as it has been used successfully on a variety of gene expression data sets [8], [10]. Given the number of clusters to be discovered, the k-means algorithm randomly selects k genes as initial cluster centroids. Depending on how far away a gene is to each centroid (using the distance/similarity measure such as the Euclidean distance [1]), it is assigned to the cluster that it is nearest to. The centroid of each cluster is then re-calculated as the mean of all genes belonging to the cluster. This process of assigning genes to the nearest clusters and re-calculating the positions of the centroids is then performed iteratively until the positions of the centroids remain unchanged.

Phase 2: A Re-clustering Phase
In this phase, genes that have already been assigned to the initial clusters (in Phase 1) are re-evaluated to determine if they should remain in the same cluster or be assigned to more than one. This phase consists of two steps as follows.

Step 1 - Discovering interesting patterns in the initial clusters
To minimize the effect of noise in this re-clustering phase, rather than the actual expression values, they are partitioned up into intervals/levels instead. The partitioning, which

is also called discretization, is based on the popular method described in [15] as this method has been shown to be able to minimize the loss of information during the data partitioning process. After discretization, interesting patterns are discovered in each initial cluster by first detecting for associations between the expression levels of genes that belong to a particular cluster and the cluster label itself. To do so, we let obs_{pk} be the observed total number of genes, $g_1,..., g_i,..., g_l$, in the data that belong to a given cluster, C_p, and are characterized by the expression values that are within the range of $e_j^{(k)}$, where $l \leq N$, $p = 1,..., P$ and P is the total number of initial clusters discovered, $e_j^{(k)}$ represents a particular level so that the expression values, $e_{1j},..., e_{ij},..., e_{lj}$, are within the range of $e_j^{(k)}$, $k = 1,..., K$ and K is the total number of distinct levels observed under E_j. We also let $\exp_{pk} = \dfrac{obs_{p+}obs_{+k}}{N^{'}}$ be the expected total under the assumption that being a member of C_p is independent of whether or not a gene has the characteristic $e_j^{(k)}$, where $obs_{p+} = \sum_{k=1}^{K} obs_{pk}$, $obs_{+k} = \sum_{p=1}^{P} obs_{pk}$ and $N^{'} = \sum_{p,k} obs_{pk} \leq N$ due to possible missing values in the data. Given obs_{pk} and $\exp_{pk}$, we are interested in determining whether obs_{pk} is significantly different from $\exp_{pk}$. To determine if this is the case, the standardized residual [16], [17] is used to scale the difference as below:

$$z_{pk} = \frac{obs_{pk} - \exp_{pk}}{\sqrt{\exp_{pk}}}. \tag{1}$$

As the above statistic approximates the standard normal distribution only when the asymptotic variance of z_{pk} is close to one. Therefore, it is, in practice, adjusted by its variance for a more precise analysis. The new test statistic, which is called the adjusted residual, can be expressed as follows:

$$d_{pk} = \frac{(obs_{pk} - \exp_{pk})/\sqrt{\exp_{pk}}}{\sqrt{v_{pk}}} = \frac{z_{pk}}{\sqrt{v_{pk}}}, \tag{2}$$

where the maximum likelihood estimate of its asymptotic variance [17], v_{pk}, is defined by

$$v_{pk} = \left(1 - \frac{obs_{p+}}{N^{'}}\right)\left(1 - \frac{obs_{+k}}{N^{'}}\right). \tag{3}$$

This statistic, Eq. (2), has an approximate standard normal distribution [18], [19], [20] and we can determine whether the expression level, $e_j^{(k)}$, under E_j is associated with C_p, at a 95% confidence level. If it is the case, $e_j^{(k)}$ is useful for determining if a gene should be grouped into C_p and the association between $e_j^{(k)}$ and C_p is statistically significant and such association is referred to as an interesting pattern.

Step 2 – Discovering overlapping clusters by the re-assignment of genes

As mentioned in Step one, we can determine whether $e_j^{(k)}$ is significantly associated with C_p using Eq. (2). If it is the case, then it can be utilized to construct characteristic description of C_p. Since such an association is not completely deterministic and the uncertainty associated with it is therefore quantified using a measure defined so that if the expression value of a gene under E_j is within the range of $e_j^{(k)}$, then it is with certainty $W(Cluster = C_p \, / \, Cluster \neq C_p \, | \, e_j^{(k)})$ that the gene belongs to C_p, where W, the weight of evidence measure [16], is defined in terms of the mutual information $I(C_p : e_j^{(k)})$ as follows:

$$W(Cluster = C_p \, / \, Cluster \neq C_p \, | \, e_j^{(k)})$$
$$= I(C_p : e_j^{(k)}) - I(\neq C_p : e_j^{(k)}) \tag{4}$$

where

$$I(C_p : e_j^{(k)}) = \log \frac{P(C_p \, | \, e_j^{(k)})}{P(C_p)} .$$

$I(C_p : e_j^{(k)})$ intuitively measures the decrease (if positive) or increase (if negative) in uncertainty about the assignment of a gene to C_p given that the expression value of this gene is within the range of $e_j^{(k)}$. Similarly, $I(\neq C_p : e_j^{(k)})$ measures the decrease (if positive) or increase (if negative) in uncertainty about the assignment of a gene to other cluster, which is not C_p, given that the expression value of this gene is within the range of $e_j^{(k)}$. Based on the mutual information, the weight of evidence, W, can be interpreted as a measure of the different in the gain in information. The weight of evidence is positive if $e_j^{(k)}$ provides positive evidence supporting the assignment of a gene to C_p; otherwise, it is negative.

To re-evaluate the cluster membership of a gene, g_i, characterized by $E_1, ..., E_j, ..., E_M$, its description can be matched against the discovered patterns. If the expression value, e_{ij}, of g_i satisfies the antecedent of a discovered pattern (e.g., the expression value, e_{ij}, is within the range of $e_j^{(k)}$) that implies C_p, then we can conclude that the description of g_i partially matches that of C_p. By repeating the above procedure, that is, by matching each expression value, e_{ij}, $j = 1, ..., M$, of g_i against the discovered patterns, the total weight of evidence of assigning g_i to C_p can be determined. Suppose that of the M characteristics that describe g_i, only M', $M' \leq M$, of them are found to match with one or more discovered patterns. Then, the total weight of evidence (*TW*) supporting the assignment of g_i to C_p is defined as follows:

$$TW(Cluster = C_p / Cluster \neq C_p \mid e_1^{(k)} ... e_j^{(k)} ... e_{M'}^{(k)})$$
$$= \sum_{j=1}^{M'} W(Cluster = C_p / Cluster \neq C_p \mid e_j^{(k)}) \tag{5}$$

Based on Eq. (5), the total weight of evidence supporting a gene belongs to each cluster, C_p, can be calculated. This measure facilitates the discovery of overlapping clusters by assigning a gene to more than one cluster only if there is a positive total weight of evidence (*TW*) of this gene to the given cluster. Moreover, it can also facilitate the identification of groups of genes (with large total weight of evidence) that have a strong association to the cluster for further biological analysis, for example, functional annotations [21]. Since, the re-clustering technique described above allows for probabilistic patterns to be detected. It performs its task by distinguishing between relevant and irrelevant expression data and by doing so, it takes into consideration global information contained in a specific cluster arrangement by evaluating the importance of different expression levels in determining cluster membership. This feature makes the proposed approach more robust to noisy data when compared to those existing algorithms that only rely on local pairwise similarity measures. In addition, during the re-clustering process, a set of interesting patterns is discovered, and each pattern can be made explicit for easy interpretation.

3   Experimental Results

3.1   Data Sets

For experimentation, we used a set of simulated data (SD1) consisting of 300 records each characterized by 50 different attributes that takes on values from [0.0, 1.0]. These records were first grouped into 3 clusters based on embedding the patterns unique to each cluster. To do so, for each cluster, we randomly selected 20% of the attributes. For each selected attribute, its values in 40% of the records in this cluster were generated randomly from within the range $[L_c, U_c]$, where L_c and U_c were also generated randomly so that $0.0 \leq L_c < U_c \leq 1.0$. To ensure that overlapping of clusters, three sets of overlapping patterns were embedded into the data as follows. Firstly, for each set of overlapping patterns, 10% of the attributes and 20% of the records were randomly selected from the whole data set. The value of a selected attribute in each selected record was then generated randomly only from within the same range $[L_f, U_f]$, where L_f and U_f were also randomly generated and $0.0 \leq L_f < U_f \leq 1.0$. In addition to simulated data, we have also tested the proposed approach using two sets of gene expression data. The first set (ED1) contains the data of 517 genes whose expression levels vary in response to serum concentration in human fibroblasts [22]. The second data set (ED2) contains the expression profiles of a subset of 384 genes which were obtained under different experimental conditions [23].

Since the total numbers of clusters in the real expression data sets are not known in advance, we therefore adopted the popular algorithm called CLICK [24] that combines graph-theoretic and statistical techniques to estimate the number of clusters, and has been widely used in gene expression data analysis. As reported by CLICK, the total number of clusters discovered in each data set is as follows: 3 clusters for SD1, 4 clusters for ED1, and 6 clusters for ED2.

The performance of the proposed approach was evaluated using the well-known silhouette measure [25]. The silhouette measure calculates the silhouette value of a record, g_i, which reflects the likelihood of g_i belonging to a cluster C_p. It does so by first estimating two scalars $a(g_i)$ and $b(g_i)$. $a(g_i)$ is the average distance between g_i and all other genes in C_p, and $b(g_i)$ is the smallest of $d(g_i, C_q)$, where $d(g_i, C_q)$ is defined to be the average distance of g_i to all genes in C_q, where $C_p \neq C_q$. The silhouette $s(g_i)$ of g_i is then defined to be the ratio,

$$\frac{b(g_i) - a(g_i)}{\max\{a(g_i), b(g_i)\}}.$$ The silhouette value lies between -1 to 1. When its value is less than zero, the corresponding gene is poorly classified. The overall silhouette value of a cluster is the average of $s(g_i)$ of all the genes in the cluster.

3.2 The Results

A. Simulated Data Set
For performance evaluation, we compared the performance of the proposed approach with the hierarchical agglomerative clustering algorithm [7], the k-means algorithm [8], SOM [9] and the fuzzy k-means algorithm [12]. It should be noted that since there is no standard measure that can be used to compare the performances between the traditional clustering algorithms and the biclustering algorithms [14], we therefore did not perform

this comparison in our study. In our experiments with the fuzzy k-means algorithm, the fuzziness parameter, m, used in our experiment was set to different values ranging from 1.1 to 2 [11] and the m that gave us the best clustering result was selected. In the simulated data, when m was set to 1.1, it gave us the best result in terms of the silhouette measure. Table 1 shows the results we obtained using the simulated data. In order to demonstrate the effectiveness of the proposed re-clustering phase in discovering overlapping clusters, we have also applied it to the hierarchical agglomerative clustering algorithm and SOM by using them separately, instead of the k-means algorithm, in the first initialization phase (note: it was not applied to the fuzzy k-means algorithms as it is able to discover overlapping clusters). By repeating the same experiments with the same set of data, the performances of these re-clustered algorithms are given in Table 2. According to the experimental results, we found that the proposed approach is rather robust in the presence of a very noisy environment. It is able to perform better than other popular clustering algorithms. When applying the proposed re-clustering phase to different clustering algorithms, it can also improve their performances.

Table 1. Comparison of the average silhouette value (SD1 data set)

Algorithm	Avg. Silhouette Value
Proposed	0.67
Hierarchical	0.45
k-means	0.48
SOM	0.5
Fuzzy k-means	0.59

Table 2. Comparison of the average silhouette value of the proposed approach using different clustering algorithms in the cluster initialization phase (SD1 data set)

Algorithm	Avg. Silhouette Value
Re-clustered Hierarchical	0.58
Re-clustered SOM	0.61

B. Gene Expression Data Sets
Based on the initial clusters discovered in Phase 1, the second re-clustering phase was performed. As described before, the re-clustering phase consists of two steps in turn. In Step one, interesting patterns were discovered in each initial cluster. Table 3 below shows some of the patterns discovered from each real expression data set that can be made explicit for easy interpretation. The patterns are expressed in rules of the form "If E_x = [L, U], then C_p [0.95]" where it should be understood as "If the expression value of a gene under E_x is within the interval from L to U, then there is a probability of 0.95 that it belongs to cluster C_p .".

Table 3. Some interesting patterns discovered from gene expression data sets (ED1 and ED2)

ED1	
If E_3 = [1.72, 2.86] then C_1 [0.82]	If E_6 = [-1.98, -0.32] then C_2 [0.71]
If E_9 = [-3.06, -1.27] then C_1 [0.79]	If E_{11} = [1.05, 3.28] then C_4 [0.86]
ED2	
If E_1 = [0.78, 2.93] then C_4 [0.8]	If E_7 = [-0.91, 1.18] then C_6 [0.9]
If E_{13} = [-1.45, -0.12] then C_2 [0.88]	If E_{16} = [1.23, 2.81] then C_3 [0.72]

Table 4. Comparison of the average silhouette value (ED1 data set)

Algorithm	Avg. Silhouette Value
Proposed	0.46
Hierarchical	0.35
k-means	0.38
SOM	0.38
Fuzzy k-means	0.42

Table 5. Comparison of the average silhouette value (ED2 data set)

Algorithm	Avg. Silhouette Value
Proposed	0.43
Hierarchical	0.31
k-means	0.36
SOM	0.34
Fuzzy k-means	0.4

Table 6. Comparison of the average silhouette value of the proposed approach using different clustering algorithms in the cluster initialization phase (ED1 data set)

Algorithm	Avg. Silhouette Value
Re-clustered Hierarchical	0.4
Re-clustered SOM	0.43

Table 7. Comparison of the average silhouette value of the proposed approach using different clustering algorithms in the cluster initialization phase (ED2 data set)

Algorithm	Avg. Silhouette Value
Re-clustered Hierarchical	0.38
Re-clustered SOM	0.4

Based on the patterns discovered, the cluster membership of each gene in each initial cluster was re-evaluated to determine if it should remain in the same cluster or be assigned to more than one. For comparison, we also compared the proposed clustering approach with other clustering algorithms as performed before. For the fuzzy k-means algorithm, the fuzziness parameter (m) was set to 1.3 for ED1 and 1.2 for ED2. As shown in Tables 4 and 5, the proposed approach still performs better than other clustering algorithms. Similar as before, we have applied the proposed re-clustering phase to the hierarchical agglomerative clustering algorithm and SOM by using them separately, instead of the k-means algorithm, in the first initialization phase. The results obtained are shown in Tables 6 and 7 below.

Since genes that have similar expression patterns may have similar or related biological functions [26], [27] and it was shown that significant enrichment of genes belonging to given functional categories can be revealed in the clusters discovered through clustering [8], therefore we also evaluated the results according to the biological functions of genes that can be discovered in each cluster. For example, when comparing the clusters discovered from the gene expression data set (ED2) after Phase 1 with those discovered after Phase 2 based on the MIPS functional catalogue database [28], we found that in each overlapping cluster, the percentage of genes in each functional category is greater than that obtained in the corresponding initial cluster (Table 8). Also, the p-value associated with each functional category discovered in the overlapping cluster is smaller than that obtained in the corresponding

Table 8. Comparison of the enrichment of genes in each functional category between the initial clusters (after Phase 1) and overlapping clusters (after Phase 2) (ED2 data set)

	MIPS Functional Category	Phase 1 (%)	Phase 1 (p-value)	Phase 2 (%)	Phase 2 (p-value)
C_1	Cell growth / morphogenesis	4.36	0.496	6.28	0.397
	Cellular sensing and response	2.51	0.403	4.79	0.28
C_2	Transported compounds	5.87	0.52	8.26	0.362
	Stress response	2.12	0.318	4.39	0.104
C_3	Cytoskeleton	10.8	0.69	19.24	0.488
	RNA synthesis	9.85	0.421	12.67	0.293
C_4	Lipid, fatty acid and isoprenoid metabolism	1.98	0.203	2.52	0.116
	Mitochondrion	13.29	0.4	25.68	0.101
C_5	Eukaryotic plasma membrane / membrane attached	3.73	0.47	6.45	0.29
	C-compound and carbohydrate metabolism	3.6	0.318	7.92	0.175
C_6	Bud / growth tip	4.83	0.58	10.19	0.43
	Fungal / microorganismic cell type differentiation	5.08	0.307	7.23	0.106

initial cluster (the *p-value* is calculated to obtain the chance probability of observing a set of genes from a particular MIPS functional category within each cluster [28], thus low *p-value* indicates high significance). According to the above results, we found that the proposed approach not only can improve the performances of existing algorithms and also can discover overlapping clusters in noisy gene expression data.

4 Conclusions

In this paper, we have presented an effective clustering approach, which consists of an initial clustering phase and a second re-clustering phase, to discover overlapping clusters in noisy gene expression data. In the initial clustering phase, local information is extracted by computing a pairwise similarity measure between gene expression profiles so as to detect for underlying clustering structures. In the second re-clustering phase, global information is obtained through discovering interesting patterns in the initial clusters. These patterns are identified by differentiating between the expression data that are relevant for the clustering process from those that are not relevant. In doing so, a probabilistic interestingness measure is used. If an expression level is relevant in determining whether or not a gene should belong to a particular cluster, it is reflected by the interestingness measure. Once the relevant expression data are identified, the discovered patterns are explicitly represented as a set of easy-to-understand if-then rules for further biological analysis. Experimental results show that the proposed approach is effective for discovering overlapping clusters in a noisy environment and also it outperforms other existing clustering algorithms. In addition, the discovered patterns, which specify the ranges of expression levels under a particular set of experimental conditions the genes should have in each cluster, may lead to further understanding of the mechanism of gene expression.

References

1. Han, J., Kamber, M.: Data Mining: Concepts and Techniques, 2nd edn. Morgan Kaufmann, San Francisco, Calif (2006)
2. Lockhart, D.J., Winzeler, E.A.: Genomic, Gene Expression and DNA Arrays. Nature 405(6788), 827–836 (2000)
3. Brazma, A., Robinson, A., Cameron, G., Ashburner, M.: One-stop Shop for Microarray Data. Nature 403(6771), 699–700 (2000)
4. Ward, J.H.: Hierarchical Grouping to Optimize an Objective Function. J. Am. Stat. Assoc. 58, 236–244 (1963)
5. MacQueen, J.: Some Methods for Classification and Analysis of Multivariate Observation. In: Proc. Symp.Math. Stat. and Prob. Berkeley., vol. 1, pp. 281–297 (1967)
6. Kohonen, T.: Self-organization and Associative Memory. Springer, New York (1989)
7. Eisen, M.B., et al.: Cluster Analysis and Display of Genome-wide Expression Patterns. Proc. Natl Acad. Sci. USA. 95(25), 14863–14868 (1998)
8. Tavazoie, S., et al.: Systematic Determination of Genetic Network Architecture. Nat. Genet. 22(3), 281–285 (1999)

9. Tamayo, P., et al.: Interpreting Patterns of Gene Expression with Self-organizing Maps: Methods and Application to Hematopoietic Differentiation. Proc. Natl. Acad. Sci. USA. 96(6), 2907–2912 (1999)
10. Berrar, D.P., Dubitzky, W., Granzow, M.: A Practical Approach to Microarray Data Analysis. Kluwer Academic Publishers, Boston Mass (2003)
11. Bezdek, J.C.: Pattern Recognition with Fuzzy Objective Function Algorithms. Plenum Press, New York (1981)
12. Gasch, A.P., Eisen, M.B.: Exploring the Conditional Coregulation of Yeast Gene Expression through Fuzzy k-means Clustering. Genome Biol 3(11), 1–22 (2002)
13. Yang, J., Wang, W., Wang, H., Yu, P.: Enhanced Biclustering on Expression Data. In: Proc. Third IEEE Conf. Bioinformatics and Bioeng, pp. 321–327 (2003)
14. Preli, A., et al.: A Systematic Comparison and Evaluation of Biclustering Methods for Gene Expression Data. Bioinformatics 22(9), 1122–1129 (2006)
15. Ching, J.Y., Wong, A.K.C., Chan, K.C.C.: Class-dependent Discretization for Inductive Learning from Continuous and Mixed-mode Data. IEEE Trans. Pattern Anal. Machine Intell. 17(7), 641–651 (1995)
16. Ewens, W.J., Grant, G.R.: Statistical Methods in Bioinformatics. Springer, Heidelberg (2005)
17. Haberman, S.J.: The Analysis of Residuals in Cross-classified Tables. Biometrics 29, 205–220 (1973)
18. Chan, K.C.C., Wong, A.K.C.: A Statistical Technique for Extracting Classificatory Knowledge from Databases. Knowledge Discovery in Databases, pp. 107–123. AAAI/MIT Press, MA (1991)
19. Au, W.H., Chan, K.C.C., Yao, X.: A Novel Evolutionary Data Mining Algorithm with Applications to Churn Modeling. IEEE Trans. Evolutionary Computation. 7(6), 532–545 (2003)
20. Chan, K.C.C., Wong, A.K.C., Chiu, D.K.Y.: Learning Sequential Patterns for Probabilistic Inductive Prediction. IEEE Trans. Systems, Man and Cybernetics 24(10), 1532–1547 (1994)
21. Mateos, A., Dopazo, J., Jansen, R., Tu, Y., Gerstein, M., Stolovitzky, G.: Systematic Learning of Gene Functional Classes from DNA Array Expression Data by Using Multiplayer Perceptrons. Genome Res 12(11), 1703–1715 (2002)
22. Iyer, V.R., et al.: The Transcriptional Program in the Response of Human Fibroblast to Serum. Science 283, 83–87 (1999)
23. Yeung, K.Y., Ruzzo, W.L.: Principal Component Analysis for Clustering Gene Expression Data. Bioinformatics 17(9), 763–774 (2001)
24. Sharan, R., et al.: CLICK and EXPANDER: A System for Clustering and Visualizing Gene Expression Data. Bioinformatics 19(14), 1787–1799 (2003)
25. Rousseeuw, J.P.: Silhouettes: A Graphical Aid to the Interpretation and Validation of Cluster Analysis. J. Comp. Appl. Math. 20, 53–65 (1987)
26. Ball, C.A., et al.: Saccharomyces Genome Database provides Tools to Survey Gene Expression and Functional Analysis Data. Nucleic Acids Res 29(1), 80–81 (2001)
27. Chu, S., et al.: The Transcriptional Program of Sporulation in Budding Yeast. Science 282, 699–705 (1998)
28. Mewes, H.W., et al.: MIPS: A Database for Genomes and Protein Sequences. Nucleic Acids Res, 31–34 (2002)

Development and Evaluation of an Open-Ended Computational Evolution System for the Genetic Analysis of Susceptibility to Common Human Diseases

Jason H. Moore, Peter C. Andrews, Nate Barney, and Bill C. White

Computational Genetics Labroatory, Department of Genetics
Dartmouth Medical School, Lebanon, NH, USA
{Jason.H.Moore,Peter.C.Andrews,Nate.Barney,Bill.C.White}@dartmouth.edu

Abstract. An important goal of human genetics is to identify DNA sequence variations that are predictive of susceptibility to common human diseases. This is a classification problem with data consisting of discrete attributes and a binary outcome. A variety of different machine learning methods based on artificial evolution have been developed and applied to modeling the relationship between genotype and phenotype. While artificial evolution approaches show promise, they are far from perfect and are only loosely based on real biological and evolutionary processes. It has recently been suggested that a new paradigm is needed where "artificial evolution" is transformed to "computational evolution" (CE) by incorporating more biological and evolutionary complexity into existing algorithms. It has been proposed that CE systems will be more likely to solve problems of interest to biologists and biomedical researchers. The goal of the present study was to develop and evaluate a prototype CE system for the analysis of human genetics data. We describe here this new open-ended CE system and provide initial results from a simulation study that suggests more complex operators result in better solutions.

1 Introduction

1.1 The Problem Domain: Human Genetics

Human genetics is undergoing an information explosion and an understanding implosion. This is the result of technical advances that make it feasible and economical to measure 10^6 or more DNA sequence variations from across the human genome. For the purposes of this paper we will focus exclusively on the single nucleotide polymorphism or SNP which is a single nucleotide or point in the DNA sequence that differs among people. Most SNPs have two alleles (e.g. A or G) that combine in the diploid human genome in one of three possible genotypes (e.g. AA, AG, GG). It is anticipated that at least one SNP occurs approximately every 100 nucleotides across the $3x10^9$ nucleotide human genome. An important goal in human genetics is to determine which of the many hundreds

E. Marchiori and J.H. Moore (Eds.): EvoBIO 2008, LNCS 4973, pp. 129–140, 2008.

of thousands of SNPs are useful for predicting who is at risk for common diseases. Further, it is important to know the nature of the mapping elationship between genotypes at the important SNPs and the phenotype or clinical endpoint. This knowledge is useful for identifying those at risk and for informing experimental studies that can lead to new therapeutic interventions.

The charge for computer science and bioinformatics is to develop algorithms for the detection and characterization of those SNPs that are predictive of human health and disease. Success in this genome-wide endeavor will be difficult due to nonlinearity in the genotype-to-phenotype mapping relationship that is due, in part, to epistasis or nonadditive gene-gene interactions. Epistasis was recognized by Bateson [1] nearly 100 years ago as playing an important role in the mapping between genotype and phenotype. Today, this idea prevails and epistasis is believed to be a ubiquitous component of the genetic architecture of common human diseases [2]. As a result, the identification of genes with genotypes that confer an increased susceptibility to a common disease will require a research strategy that embraces, rather than ignores, this complexity [2,3,4]. The implication of epistasis from a data mining point of view is that SNPs need to be considered jointly in learning algorithms rather than individually. Because the mapping between the attributes and class is nonlinear, the concept difficulty is high. The challenge of modeling attribute interactions has been previously described [5]. The goal of the present study is to develop an evolutionary computing strategy for detecting and characterizing epistasis.

1.2   A Simple Example of the Concept Difficulty

Epistasis can be defined as biological or statistical [3]. Biological epistasis occurs at the cellular level when two or more biomolecules physically interact. In contrast, statistical epistasis occurs at the population level and is characterized by deviation from additivity in a linear mathematical model. Consider the following simple example of statistical epistasis in the form of a penetrance function. Penetrance is simply the probability (P) of disease (D) given a particular combination of genotypes (G) that was inherited (i.e. $P[D|G]$). A single genotype is determined by one allele (i.e. a specific DNA sequence state) inherited from the mother and one allele inherited from the father. For most single nucleotide polymorphisms or SNPs, only two alleles (encoded by A or a) exist in the biological population. Therefore, because the order of the alleles is unimportant, a genotype can have one of three values: AA, Aa or aa. The model illustrated in Table 1 is an extreme example of epistasis. Let's assume that genotypes AA, aa, BB, and bb have population frequencies of 0.25 while genotypes Aa and Bb have frequencies of 0.5 (values in parentheses in Table 1). What makes this model interesting is that disease risk is dependent on the particular combination of genotypes inherited. Individuals have a very high risk of disease if they inherit Aa or Bb but not both (i.e. the exclusive OR function). The penetrance for each individual genotype in this model is 0.5 and is computed by summing the products of the genotype frequencies and penetrance values. Thus, in this model there is no difference in disease risk for each single genotype as specified

Table 1. Penetrance values for genotypes from two SNPs

	AA (0.25)	Aa (0.50)	aa (0.25)
BB (0.25)	0	1	0
Bb (0.50)	1	0	1
bb (0.25)	0	1	0

by the single-genotype penetrance values. This model was first described by Li and Reich [6]. Heritability, or the size of the genetic effect, is a function of these penetrance values. In this model, the heritability is maximal at 1.0 because the probability of disease is completely determined by the genotypes at these two DNA sequence variations. As Freitas [5] reviews, this general class of problems has high concept difficulty.

1.3 Towards Computational Evolution for the Analysis of Gene-Gene Interactions

Numerous machine learning and data mining methods have been developed and applied to the detection of gene-gene interactions. These include, for example, traditional methods such as neural networks [7] and novel methods such as multifactor dimensionality reduction [8]. Evolutionary computing methods such as genetic programming (GP) have been applied to both attribute selection and model discovery in the domain of human genetics. For example, Ritchie et al. [8] used GP to optimize both the weights and the architecture of a neural network for modeling gene-gene interactions. More recently, GP has been successfully used for both attribute selection [9,10] and genetic model discovery [11].

Genetic programming is an automated computational discovery tool that is inspired by Darwinian evolution and natural selection [12,13,14,15,16,17,18]. The goal of GP is evolve computer programs to solve problems. This is accomplished by first generating random computer programs that are composed of the building blocks needed to solve or approximate a solution to a problem. Each randomly generated program is evaluated and the good programs are selected and recombined to form new computer programs. This process of selection and recombination is repeated until a best program is identified.

Genetic programming has been applied successfully to a wide range of different problems including data mining and knowledge discovery [e.g. [19]] and bioinformatics [e.g. [20]]. Despite the many successes, there are a large number of challenges that GP practitioners and theorists must address before this general computational discovery tool becomes one of several tools that a modern problem solver calls upon [21]. Banzhaf et al. [22] propose that overly simplistic and abstracted artificial evolution (AE) methods such as GP need to be transformed into computational evolution (CE) systems that more closely resemble the complexity of real biological and evolutionary systems. Evolution by natural selection solves problems by building complexity. As such, computational systems inspired by evolution should do the same. The working hypothesis addressed in the present study is that a GP based genetic analysis system will find better solutions faster if it is implemented as a CE system that can evolve a variety of complex operators that in turn generate variability in solutions. This is in contrast to an AE system that uses a fixed set of operators.

1.4 Research Questions Addressed and Overview

The goal of the present study was to develop and evaluate an open-ended CE system for the detection and characterization of epistasis. We developed a hierarchically-organized and spatially-extended GP approach that is capable of evolving its own operators of any arbitrary size and complexity. The primary question addressed in this study is whether the ability to evolve complex operators improves the ability of the system to discover a classifier that is capable of predicting disease in the presence of nonlinear gene-gene interactions.

2   A Prototype Computational Evolution System

Our primary goal was to develop a prototype CE system that is capable of open-ended evolution for bioinformatics problem-solving in the domain of human genetics. Figure 1 gives a graphical overview of our hierarchically-organized and spatially-extended GP system that is capable of open-ended CE. At the bottom layer of this hierarchy is a grid of solutions. Details of the solutions and their representation are given in Section 2.1. At the second layer of the hierarchy is a grid of operators of any size and complexity that are capable of modifying the solutions. The operators are described in Section 2.2. At the third layer in the hierarchy is a grid of mutation operators that are capable of modifying the solution operators. The mutation operators are described in Section 2.3. At the highest level of the hierarchy is the mutation frequency that determines the rate at which operators are mutated. This is described in Section 2.4. Details of how the system was implemented are described in Section 2.5. The details of the experimental design used to evaluate this system are described in Section 3.

2.1 Problem Solutions: Their Representation, Fitness Evaluation and Reproduction

The goal of a classifier is to accept as input two or more discrete attributes (i.e. SNPs) and produce a discrete output that can be used to assign class (i.e. healthy or sick). Here, we used symbolic discriminant analysis or SDA as our classifier. The SDA method [23] has been described previously for this problem domain [11]. Briefly, SDA models consist of a set of attributes and constants as input and a set of mathematical functions that produce for each instance in the dataset a score called a symbolic discriminant score. The goal of SDA is to find a linear or nonlinear combination of attributes such that the difference between the distributions of symbolic discriminant scores for each class is maximized. Here, our SDA function set was $\{+, -, *, /, \%, <, <=, >, >=, ==, \neq\}$ where the % operator is a mod operation and / is a protected division. The SDA models are represented as postfix expressions here instead of as expression trees as has been used in the past [23,11] to facilitate stack-based evaluation of the classifiers and to facilitate representation in text files.

Classification of instances into one of the two classes requires a decision rule that is based on the symbolic discriminant score. Thus, for any given symbolic discriminant score (S_{ij}) in the ith class and for the jth instance, a decision rule can be formed such that if $S_{ij} > S_o$ then assign the instance to one class and if $S_{ij} <= S_o$ then assign the observation to the other class. When the prior probability that an instance belongs to one class is equal to the probability that it belongs to the other class, S_o can be defined as the arithmetic mean of the median symbolic discriminant scores from each of the two classes. This is the classification rule we used in the present study and is consistent with previous work in this domain [11]. Using this decision rule, the classification accuracy for a particular discriminant function can be estimated from the observed data. Here, accuracy is defined as $(TP+TN)/(TP+TN+FP+FN)$ where TP are true positives (TP), TN are true negatives, FP are false positives, and FN are

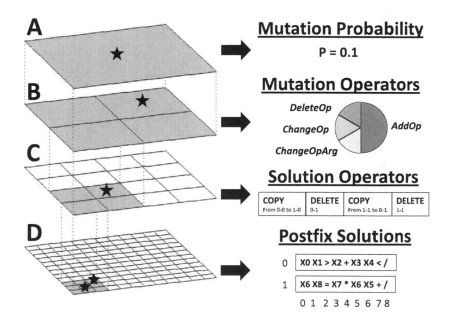

Fig. 1. Visual overview of our prototype CE system. The hierarchical structure is shown on the left while some specific examples at each level are shown on the right. The top two levels of the hierarchy (A and B) exist to generate variability in the operators that modify the solutions. Shown in C is an example set of operators that will perform recombination on the two solutions shown in D. As illustrated in B, there is a 0.50 probability that a mutation to the recombination operator in C will add an operator thus making this particular operator more complex. This system allows operators of any arbitrary complexity to modify solutions. Note that we used a 24x24 grid of solutions in the present study. A 12x12 grid is shown as an illustrative example.

false negatives. We used accuracy as the fitness measure for SDA solutions as has been described previously [11].

All SDA solutions in a population are organized on a toroidal grid with specific X and Y coordinates (see example in Figure 1). As such, they resemble previous work on cellular genetic programming [24]. In the present study we used a 24x24 grid for a total population size of 576. Reproduction of solutions in the population is handled in a spatial manner. Each solution is considered for reproduction in the context of its Moore neighborhood using an elitist strategy. That is, each solution in question will compete with its eight neighbors and be replaced in the next generation by the neighbor with the highest fitness of all solutions. This combines ideas of tournament selection that is common in GP with a set of solutions on a grid. Variability in solutions is generated using hierarchically organized operators. This is described below.

2.2   Operators for Computational Evolution: Generating Solution Variability

Traditional AE approaches such as GP use a fixed set of operators that include mutation and recombination, for example. The goal of developing a prototype CE system was to provide operators and building blocks for operators that could be combined to create new operators of any arbitrary complexity. We started with the following six operators and operator building blocks. The first operator, DeleteRangeOperation, deletes all functions in an SDA postfix expression within a certain range. The second operator, CopyRangeOperator, copies all functions in an SDA postfix expression within a certain range to another SDA postfix expression at a particular position. The third operator, PermuteRange-Operator, randomizes the order of a set of SDA functions within a given range. The fourth operator, AddOperator, adds a randomly selected function onto the end of a set of SDA functions. The fifth operator, PointMutationOperator, replaces a function and its arguments (e.g. attributes) at a given position with a randomly selected function and arguments. The final operator, PointMuta-tionExpertKnowledgeOperator, replaces a function and its arguments (e.g. attributes) at a given position with a randomly selected function and arguments selected using a source of expert knowledge. Greene et al. [25] have shown that using ReliefF measures of attribute quality to guide point mutation for genetic analysis using GP is beneficial for ensuring good building blocks are utilized. This is consistent with Goldberg's ideas about exploiting good building blocks in competent genetic algorithms [26]. Thus, we have provided to the CE system a set of operators and operator building blocks that can be put together in any arbitrary length and complexity. For example, a standard recombination operator can be formed by combining two CopyRangeOperator operators and two DeleteRangeOperation operators with the appropriate arguments that specify the correct positions in two SDA solutions for copying and deleting appropriate model pieces. An example recombination operator is shown in Figure 1. These operators can be combined in more interesting ways to form even more complex operators.

As with the solutions, each operator is organized on a toroidal grid with a specific X and Y coordinate. Rather than generate one operator for each solution we assigned each operator to a set of solutions. This makes evaluation of the fitness of an operator easier since its positive or negative effect on the solutions can be averaged over multiple solutions. In this study, we assigned each operator to a 6x6 grid of 36 solutions. Thus, the population of operators is organized in a 4x4 grid for a total of 16 operators (See Figure 1) that each maps onto 36 of the 576 solutions.

2.3 Mutation of Operators for Computational Evolution: Generating Operator Variability

An important goal for the prototype CE system is the ability to generate variability in the operators that modify solutions. To accomplish this goal we developed an additional level in the hierarchy (Figure 1B) with mutation operators that specifically alter the operators described above. We defined four different fixed mutation operators that are each assigned to a 2x2 grid of solution operators. Solution operators can be modified in the following four ways. First, an operator can have a specific operator building block deleted (DeleteOperator). Second, an operator can have a specific operator building block added (AddOperator). Third, an operator can have a specific operator building block changed (ChangeOperator). Finally, an operator can have its arguments changed (ChangeOperatorArguments). This latter function allows, for example, the range that a DeleteRangeOperation would use. For our prototype, we fixed the probabilities with which each of these types of mutations can change the operators. Here, we used all four types of mutation and defined four different probability distributions for their use. For the first distribution we set the probabilities for DeleteOperator, AddOperator, ChangeOperator and ChangeOperatorArguments to 0.5, 0.167, 0.167 and 0.167 respectively. For the second distribution we set the probabilities to 0.167, 0.5, 0.167 and 0.167. For the third we set the probabilities to 0.167, 0.167, 0.5 and 0.167 and for the fourth we set the probabilities to 0.167, 0.167, 0.167 and 0.5. This preliminary assignment of probabilities allows us to explore the usefulness of each type of mutation. In future versions the type of mutation and their probabilities will also evolve. These four sets of mutations that alter solution operators exist in a 2x2 grid. Each mutates four sets of operators at the next level down in the hierarchy (see Figure 1).

2.4 Mutation Frequency

The top level of the CE system hierarchy (see Figure 1) is the mutation frequency that controls the probability that one of the four mutation sets in the next level down will mutate a given solution operator two levels down. In the present study we fixed this to 0.1. In future version this will be an evolvable parameter. Note that this frequency does not control the frequency with which an operator modifies a solution in the lowest level. This is controlled by the operator itself when it specifies which solution(s) it will modify.

2.5   Implementation

The CE system was programmed in C++. A single run of the system with a population of 576 solutions on a 24x24 grid for 100 generations took approximately three minutes on an 2.2 GHz AMD Opteron processor.

3   Experimental Design

Our goal was to provide an initial evaluation of the prototype CE system described above. The central question addressed in this study is whether the ability to evolve operators of any arbitrary complexity improves quality of the SDA models. To address this question, we first ran, as a baseline, the CE system that utilized only a simple mutation operator. Next, we ran the CE system with all available operators. Each run was completed with a population size of 576 (24x24 solutions) for 100 generations and 1000 generations. The best model from each run was saved along with the accuracy of the symbolic discriminant function. Each method was run 100 times with different random seeds on data that was simulated using the penetrance function in Table 2 below. The data consisted of 1600 instances and two functional SNPs that are associated with class only through the type of nonlinear interaction described in Section 1.2. The heritability of this model is 0.4. Each dataset also consisted of 98 randomly generated SNPs that represent potential false-positives or noise in the data. The challenge for the CE system is to search for the right combination of two SNPs and identify a nonlinear function that approximates the pattern generated by the penetrance model in Table 2. It is important to note that target classification accuracy for the correct model is approximately 0.8.

Table 2. Penetrance values for genotypes from two SNPs used to simulate data

	AA (0.04)	Aa (0.32)	aa (0.64)
BB (0.04)	0.486	0.960	0.538
Bb (0.32)	0.947	0.004	0.811
bb (0.64)	0.640	0.606	0.908

The distribution of accuracies obtained from running the CE system with just a simple mutation operator versus running the system with the capability of generating more complex operators were statistically compared using a two-sample t-test. The two systems were considered statistically significant at a type I error rate of 0.05.

4   Results

Figure 2 below summarizes the distribution of accuracies obtained from running the CE system 100 times on the simulated data with evolved operators (All)

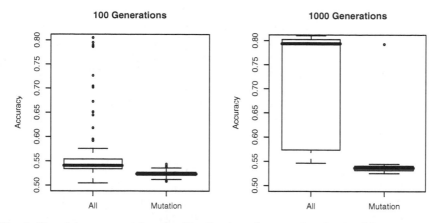

Fig. 2. Boxplots summarizing the distribution of accuracies obtained from running the CE system 100 times on the simulated data with evolved operators (All) or with just a mutation perator (Mutation) for 100 generations and 1000 generations

or with just a mutation operator (Mutation) for 100 generations and 1000 generations. The line in the middle of each box is the median of the distribution while the upper and lower limits of the box itself represent the 25th and 75th percentiles. The dashed lines extending from each box represent the approximate range of values with circles representing extreme values. Note that at 1000 generations, mutation alone only approximated the correct answer once out of 100 runs while the full CE system approximated the correct answer more than 50% of the time. In both cases, the mean accuracy was significantly higher for the full system ($P < 0.05$).

These preliminary results indicate that, for this specific domain, a CE system with the ability to evolve operators of any size and complexity does indeed identify better solutions than a baseline system that uses a fixed mutation operator. An important question is whether more complex operators were actually used to generate the best models discovered by the CE system. We evaluated the operators discovered during each run that were associated with a best model and found that all six operators and operator building blocks defined in Section 2.2 were used at least once in each of the 100 runs. This demonstrates that complex operators were discovered and used to generate better solutions than a simple mutation operator was able to generate.

5   Discussion and Conclusions

Banzhaf et al. [22] have suggested that traditional artificial evolution methods such as genetic programming (GP) will greatly benefit from our current understanding of the complexity of biological and evolutionary systems. They propose a new research agenda in which CE systems that mimic the complexity of biological systems will replace the overly simplified artificial evolution systems that have been inspired by biology, but largely ignore the complexity of biological processes. The goal of the present study was to specifically address whether a computational evolution system capable of evolving more complex operators will find better solutions than an artificial evolution system in the domain of human genetics. To accomplish this goal we developed a prototype CE system that is both spatially and hierarchically organized and is capable of evolving operators of any arbitrary size and complexity from a set of basic operator building blocks. Our preliminary experimental results demonstrate that the ability to evolve more complex operators does indeed improve the ability of the system to identify good models. These results support our working hypothesis and are consistent with the research agenda proposed by Banzhaf et al. [22].

It is important to note that the system presented here is a prototype and, as such, there are many extensions and modifications that can be made that would be consistent with CE. We first discuss several features implemented in the prototype that add complexity to the system and then propose some additional features inspired by the complexity of biological systems. There were two primary sources of complexity. First, the system is capable of evolving a diversity of different operators that modify solutions in the spatially-organized population. This is similar to real biological systems that evolve more complex genomic processes. For example, microRNAs that participate in post-translational regulation have evolved, in part, to help determine developmental processes such as body plan specification. Sempere et al. [27] showed that the number of microRNAs an animal group has correlates strongly with the hierarchy of metazoan relationships. The ability of species to evolve new biological processes plays an important role in increasing their complexity. As a second feature, we have included in the set of operator building blocks a mutation function that responds to the environment (i.e. the expert knowledge). We know that expert knowledge in the form of other data mining results or biological information about gene function is critical for success in this domain [9,10,11]. Here, we gave the CE operators the ability to use expert knowledge (i.e. information from the environment) in the form of pre-processed ReliefF scores to preferentially choose good attributes as arguments for a new function. The ability of an organism to respond to its environment plays an important role in fitness. The important role of environmental sensing has been discussed [22].

Our future goal is to improve the prototype CE system by adding additional features that are inspired by the complexity of real biological systems. As a first step, we will make the mutation operators (see Section 2.3, Figure 1B) more complex by giving them the ability to evolve. That is, the probability distribution that controls how operators are modified through mutation will evolve with feedback from how the system is doing. We also make the overall mutation frequency at the highest level an evolvable parameter. The evolvability of the entire system will make it attractive to implement this system in parallel as an island model thus providing a virtual ecosystem with feedback between populations. As a second step, we will add additional feedback loops in the system. For example, the solutions could contribute information back to the environment that is then used by a complex operator to generate variability in solutions. This takes the environmental sensing idea discussed above a step further. We anticipate the addition of these types of feedback loops will significantly increase the complexity of the system. Whether these additional features con tinues to improve the ability of this machine learning method to solve complex problems in human genetics still needs to be addressed.

Acknowledgments

This work was supported by National Institutes of Health (USA) grants LM009012, AI59694 and RR018787. We thank Dr. Wolfgang Banzhaf and the attendees of the 2007 Genetic Programming Theory and Practice (GPTP) Workshop for their insightful ideas about CE and complexity.

References

1. Bateson, W.: Mendel's Principles of Heredity. Cambridge University Press, Cambridge (1909)
2. Moore, J.H.: The ubiquitous nature of epistasis in determining susceptibility to common human diseases. Human Heredity 56, 73–82 (2003)
3. Moore, J.H., Williams, S.W.: Traversing the conceptual divide between biological and statistical epistasis: Systems biology and a more modern synthesis. BioEssays 27, 637–646 (2005)
4. Thornton-Wells, T.A., Moore, J.H., Haines, J.L.: Genetics, statistics and human disease: Analytical retooling for complexity. Trends in Genetics 20, 640–647 (2004)
5. Freitas, A.: Understanding the crucial role of attribute interactions. Artificial Intelligence Review 16, 177–199 (2001)
6. Li, W., Reich, J.: A complete enumeration and classification of two-locus disease models. Human Heredity 50, 334–349 (2000)
7. Lucek, P.R., Ott, J.: Neural network analysis of complex traits. Genetic Epidemiology 14, 1101–1106 (1997)
8. Ritchie, M.D., Hahn, L.W., Roodi, N., Bailey, L.R., Dupont, W.D., Parl, F.F., Moore, J.H.: Multifactor dimensionality reduction reveals high-order interactions among estrogen metabolism genes in sporadic breast cancer. American Journal of Human Genetics 69, 138–147 (2001)
9. Moore, J.H., White, B.C.: Genome-wide genetic analysis using genetic programming: The critical need for expert knowledge. In: Riolo, R.L., Soule, T., Worzel, B. (eds.) Genetic Programming Theory and Practice IV. Genetic and Evolutionary Computation, vol. 5, Springer, Heidelberg (2006)
10. Moore, J.H., White, B.C.: Exploiting expert knowledge in genetic programming for genome-wide genetic analysis. In: Runarsson, T.P., Beyer, H.-G., Burke, E.K., Merelo-Guervós, J.J., Whitley, L.D., Yao, X. (eds.) PPSN 2006. LNCS, vol. 4193, pp. 969–977. Springer, Heidelberg (2006)
11. Moore, J.H., Barney, N., Tsai, C.T., Chiang, F.T., Gui, J., White, B.C.: Symbolic modeling of epistasis. Human Heredity 63(2), 120–133 (2007)
12. Koza, J.R.: Genetic Programming: On the Programming of Computers by Means of Natural Selection. MIT Press, Cambridge (1992)

13. Koza, J.R.: Genetic Programming II: Automatic Discovery of Reusable Programs. MIT Press, Cambridge Massachusetts (1994)
14. Koza, J.R., Andre, D., Bennett, I.F.H., Keane, M.: Genetic Programming 3: Darwinian Invention and Problem Solving. Morgan Kaufmann, San Francisco (1999)
15. Koza, J.R., Keane, M.A., Streeter, M.J., Mydlowec, W., Yu, J., Lanza, G.: Genetic Programming IV: Routine Human-Competitive Machine Intelligence. Kluwer Academic Publishers, Dordrecht (2003)
16. Banzhaf, W., Nordin, P., Keller, R.E., Francone, F.D.: Genetic Programming – An Introduction. In: On the Automatic Evolution of Computer Programs and its Applications. Morgan Kaufmann, San Francisco (January, 1998)
17. Langdon, W.B.: Genetic Programming and Data Structures: Genetic Programming + Data Structures = Automatic Programming!, Genetic Programming, vol. 1. Kluwer, Boston (April 24, 1998)
18. Langdon, W.B., Poli, R.: Foundations of Genetic Programming. Springer, Heidelberg (2002)
19. Freitas, A.: Data Mining and Knowledge Discovery with Evolutionary Algorithms. Springer, Heidelberg (2002)
20. Fogel, G.B., Corne, D.W.: Evolutionary Computation in Bioinformatics. Kaufmann Publishers, San Francisco (2003)
21. Yu, T., Riolo, R.L., Worzel, B. (eds.): Genetic Programming Theory and Practice III. Genetic Programming, vol. 9. Ann Arbor, Springer, Heidelberg (May 12–14, 2005)
22. Banzhaf, W., Beslon, G., Christensen, S., Foster, J.A., Kepes, F., Lefort, V., Miller, J., Radman, M., Ramsden, J.J.: From artificial evolution to computational evolution: a research agenda. Nature Reviews Genetics 7, 729–735 (2006)
23. Moore, J.H., Parker, J.S., Hahn, L.W.: Symbolic discriminant analysis for mining gene expression patterns. In: Flach, P.A., De Raedt, L. (eds.) ECML 2001. LNCS (LNAI), vol. 2167, pp. 191–205. Springer, Heidelberg (2001)
24. Folino, G., Pizzuti, C., Spezzano, G.: A cellular genetic programming approach to classification. In: Proceedings of the Genetic and Evolutionary Computation Conference (GECCO 1999), pp. 1015–1020 (1999)
25. Greene, C.S., White, B.C., Moore, J.H.: An expert knowledge-guided mutation operator for genome-wide genetic analysis using genetic programming, vol. 4774, pp. 30–40 (2007)
26. Goldberg, D.E.: The Design of Innovation. Kluwer Academic Publishers, Dordrecht (2002)
27. Sempere, L.F., Cole, C.N., McPeek, M.A., Peterson, K.J.: The phylogenetic distribution of metazoan micrornas: insights into evolutionary complexity and constraint. Journal of Experimental Zoology 306, 575–575 (2006)

Gene Selection and Cancer Microarray Data Classification Via Mixed-Integer Optimization

Carlotta Orsenigo

Dip. di Scienze Economiche, Aziendali e Statistiche, Università di Milano, Italy
`carlotta.orsenigo@unimi.it`

Abstract. The growing availability of biological measurements at the molecular level has recently enhanced the role of machine learning methods for effective early cancer diagnosis, prognosis and treatment. These measurements are represented by the expression levels of thousands of genes in normal and tumor sample tissues. In this paper we present a two-phase algorithm for gene expression data classification. In the first phase, a novel gene selection method based on mixed-integer optimization is applied with the aim of selecting a small subset of cancer marker genes. In the second phase, a binary polyhedral classifier is used in order to label gene expression data. Computational experiments performed on three benchmark datasets indicate the usefulness of the proposed framework which is capable of competitive performances with respect to the best classification accuracy so far achieved for each dataset. Moreover, the classification rules generated are based on very few genes which, in our computations, can be credited as the most influential genes for tumor differentiation.

Keywords: Gene selection, microarray data classification, mixed-integer optimization, discrete support vector machines.

1 Introduction

Cancer is a family of diseases originated from genetic abnormalities which may be inherited, due to errors in the DNA replication or caused by the prolonged exposure to promoting agents, such as tobacco smoke, chemicals and radiation. The classification of a cancer is essential for administering the most effective treatment, and has been traditionally based on the analysis of its morphological appearance. However, tumors sharing similar histopathological features, and thereby assigned to the same diagnostic category, may follow different clinical courses and respond differently to the therapy assigned. Furthermore, the genetical nature of cancer alterations calls for the development of new classification methods based on detailed gene expression data, in order to fully exploit cancer molecular characteristics (Cuperlovic-Culf et al., 2005).

The advances in microarray technology and the growing availability of biological measurements performed at the molecular level have intensified the role of machine learning methods for effective cancer classification and diagnosis.

E. Marchiori and J.H. Moore (Eds.): EvoBIO 2008, LNCS 4973, pp. 141–152, 2008.

These measurements are represented by the expression levels of thousands of genes exhibited in tumor and normal tissues under the same experimental conditions. The collection of gene expression data usually results in high dimensional datasets, composed by a huge number of features (genes) and a relative few number of sample tissues. As a consequence, the established paradigm for microarray data classification consists of a two-phase procedure in which gene selection is performed as a separate processing step in order to avoid the curse of dimensionality which is likely to be encountered by the majority of classification methods (Bishop, 1995). In particular, in the first phase the most informative features, that is genes whose expression profiles characterize a particular state, are identified. Then, the classification of gene expression patterns is performed using the features selected in the first phase (Wang et al., 2007; Shah and Kusiak, 2007; Wong and Hsu, 2008).

Several gene selection methods have been proposed in order to improve the efficiency of microarray data classification, and possibly increase its accuracy. Most of these methods are univariate in nature, since they provide a ranking of the genes considered individually, disregarding their correlations. Different rankings may be obtained according to alternative evaluation criteria, ranging from signal-to-noise ratio (Golub et al., 1999), to information gain (Yang et al., 2003) and gene regulation probabilities (Wang et al., 2007). Other approaches rely on multivariate analysis, with the aim of studying the relevance of multiple genes simultaneously and capturing their correlations. The reader may refer to (Lai et al., 2006) for a comprehensive comparison of gene selection methods for cancer classification. A wide variety of machine learning algorithms have been proposed also for microarray data classification, such as k-nearest neighbor (Bagui et al., 2003), Bayesian classifiers (Mallick et al., 2005), neural networks (Cho and Won, 2007) and support vector machines (Furey et al., 2000).

In this paper we propose a novel two-phase technique for cancer microarray data classification. In the first phase, a new gene selection method based on mixed-integer optimization is applied with the aim of identifying a small subset of genes which may be useful for discriminating between normal and tumor tissues. More specifically, the optimization problem attempts to isolate a predefined number of genes whose expression levels are similar for tissues in the same class and differ for tissues of opposite classes. In the second phase, a binary polyhedral classifier, recently proposed in (Orsenigo and Vercellis, 2007), is used in order to label gene expression profiles, and derive classification rules for predicting the state of future tissues. This classifier, which is able to perform linear and non-linear separations, has proven to perform quite efficiently on small datasets, and thereby appears to be adequate for the classification of cancer microarray datasets, which are usually composed by a low number of sample tissues.

The usefulness of the proposed approach has been evaluated on an empirical basis by means of computational tests performed on three publicly available datasets. The classification performances exhibited by our method are competitive with respect to the best results so far provided in the literature. Moreover, the classification rules generated by the polyhedral classifier are rather simple, since they are based on small subsets of genes. In our computations, these genes can be credited as the most influential features for tumor differentiation.

2 Gene Selection by Mixed-Integer Optimization

In cancer microarray data classification we are provided with an expression matrix $\mathcal{S}_{m \times n}$ in which rows correspond to sample tissues of patients, and columns represent the genes whose expression has been measured under the same experimental conditions. The value of $\mathcal{S}$ at row i and column j is therefore provided by the expression level of gene j observed for patient i. Let $\mathbf{x}_i \in \mathbb{R}^n$ be the vector of expression values corresponding to row i in the matrix $\mathcal{S}$. For each example $\mathbf{x}_i, i \in \mathcal{M} = \{1, 2, \ldots, m\}$, it is known the value y_i of a categorical variable indicating the class associated to $\mathbf{x}_i$. In what follows, we will confine the attention to binary classification problems in which the class of each example $\mathbf{x}_i$ can take only two different values. In particular, without loss of generality we will set $y_i = -1$ or $y_i = 1$ if example $\mathbf{x}_i$ represents a normal or a tumor sample tissue, respectively (Fig. 1).

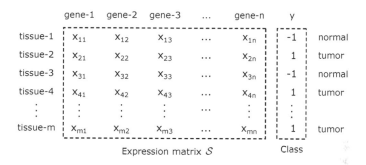

Fig. 1. Gene expression matrix and class values

Under these assumptions, solving a cancer microarray data classification problem requires to find a function $f(\mathbf{x}) : \mathbb{R}^n \mapsto \{-1, 1\}$ which optimally describes the relationship between the sample tissues and their class values, in order to predict the diagnostic category of future examples.

To the end of identifying the most relevant genes for the classification of $\mathcal{S}$ we present a novel gene selection method based on the solution of a linear mixed-integer optimization problem. The proposed method attempts to isolate genes whose profiles characterize a particular state, that is genes whose expression levels are similar for patients in the same class and differ for patients of opposite classes. For this reason, the mixed-integer optimization model represents a classification method itself, searching for the best relationship between the gene expression patterns and the class values of the sample tissues.

The selection procedure starts by organizing the gene expression profiles in the form required by the optimization problem. In particular, the original expression

(a)

s^-	gene-1	gene-2	gene-3	...	gene-n	y
tissue-1	x_{11}	x_{12}	x_{13}	...	x_{1n}	-1
tissue-3	x_{31}	x_{32}	x_{33}	...	x_{3n}	-1
⋮	⋮	⋮	⋮		⋮	⋮
s^+						
tissue-2	x_{21}	x_{22}	x_{23}	...	x_{2n}	1
tissue-4	x_{41}	x_{42}	x_{43}	...	x_{4n}	1
⋮	⋮	⋮	⋮			
tissue-m	x_{m1}	x_{m2}	x_{m3}	...	x_{mn}	1

(b)

S^-

	tissue-1	tissue-3	...	tissue-(s^-)
gene-1	x_{11}	x_{31}	...	$x_{(s^-)1}$
gene-2	x_{12}	x_{32}	...	$x_{(s^-)2}$
gene-3	x_{13}	x_{33}	...	$x_{(s^-)3}$
⋮	⋮	⋮		⋮
gene-n	x_{1n}	x_{3n}	...	$x_{(s^-)n}$
y	-1	-1	...	-1

S^+

	tissue-2	tissue-4	...	tissue-(s^+)
gene-1	x_{21}	x_{41}	...	$x_{(s^+)1}$
gene-2	x_{22}	x_{42}	...	$x_{(s^+)2}$
gene-3	x_{23}	x_{43}	...	$x_{(s^+)3}$
⋮	⋮	⋮		⋮
gene-n	x_{2n}	x_{5n}	...	$x_{(s^+)n}$
y	1	1	...	1

(c)

G	tissue-1'	tissue-2'	...	tissue-(s)	tissue-(s+1)	tissue-(s+2)	...	tissue-(2s)
gene-1	h_{11}	h_{12}	...	$h_{1(s)}$	$h_{1(s+1)}$	$h_{1(s+2)}$	...	$h_{1(2s)}$
gene-2	h_{21}	h_{22}	...	$h_{2(s)}$	$h_{2(s+1)}$	$h_{2(s+2)}$	...	$h_{2(2s)}$
gene-3	h_{31}	h_{32}	...	$h_{3(s)}$	$h_{3(s+1)}$	$h_{3(s+2)}$	...	$h_{3(2s)}$
⋮	⋮	⋮	...	⋮	⋮	⋮		⋮
gene-n	h_{n1}	h_{n2}	...	$h_{n(s)}$	$h_{n(s+1)}$	$h_{n(s+2)}$	...	$h_{n(2s)}$
y	-1	-1	...	-1	1	1	...	1

(d)

G_t	t-1	t-2	...	t-s	y
gene-1^-	$(g_{11})^-$	$(g_{12})^-$	...	$(g_{1s})^-$	-1
gene-2^-	$(g_{21})^-$	$(g_{22})^-$	...	$(g_{2s})^-$	-1
gene-5^-	$(g_{51})^-$	$(g_{52})^-$	...	$(g_{5s})^-$	-1
⋮	⋮	⋮		⋮	⋮
gene-f^-	$(g_{r1})^-$	$(g_{r1})^-$	...	$(g_{rs})^-$	-1
gene-1^+	$(g_{11})^+$	$(g_{12})^+$	...	$(g_{1s})^+$	1
gene-2^+	$(g_{21})^+$	$(g_{22})^+$	...	$(g_{2s})^+$	1
gene-5^+	$(g_{51})^+$	$(g_{52})^+$	...	$(g_{5s})^+$	1
⋮	⋮	⋮		⋮	⋮
gene-f^+	$(g_{r1})^+$	$(g_{r1})^+$	...	$(g_{rs})^+$	1

Fig. 2. First step of the gene selection procedure

matrix S is divided into two disjoint datasets, each containing examples in the same diagnostic category (Fig. 2-a). These datasets are then transposed in order to derive two new matrices, S^- and S^+, of size $(n \times s^-)$ and $(n \times s^+)$, where s^- and s^+ indicate the number of normal and tumor examples in S. Notice that in S^- and S^+ the rows represent the genes whereas the columns correspond to the patients (Fig. 2-b). In this way, S^- and S^+ collect the expression profile of each individual gene in normal and tumor tissues, respectively.

If S^- and S^+ are of different size ($s^- \neq s^+$), from each of the two matrices only $s = \min(s^-, s^+)$ columns are randomly retained. This allows to combine the columns in S^- and S^+ so as to obtain a final matrix G of size $(n \times 2s)$, taking the form described in Fig. 2-c. Finally, to reduce the computational effort required by the subsequent steps, the rows in G are randomly partitioned into r disjoint subsets $G_1, G_2, \ldots, G_r$ of approximately equal size. Each dataset $G_t, t = 1, 2, \ldots, r$, is obtained by means of an *ad-hoc* sampling procedure so that it contains for each gene the expression patterns for both the normal and the tumor sample tissues (Fig. 2-d). In order to simplify the notation, hereafter the analysis will be referred to the generic dataset G_t.

For feature selection, the optimization problem attempts to classify the genes in G_t by means of a discriminant function taking the form of a separating hyperplane. More specifically, it is based on the idea of identifying genes whose expression patterns differ significantly according to the state of the sample tissues. Let $\mathcal{F} = \{1, 2, \ldots, f\}$ denote the set of the indices of the different genes in G_t, and $\mathbf{g}_i^-$ and $\mathbf{g}_i^+, i \in \mathcal{F}$, the vectors of the expression levels of gene i for normal and tumor tissues, respectively. For each gene $i \in \mathcal{F}$, define two binary variables

$$z_i^- = \begin{cases} 0 & \text{if the profile } \mathbf{g}_i^- \text{ properly identifies normal tissues} \\ 1 & \text{otherwise} \end{cases}, \tag{1}$$

$$z_i^+ = \begin{cases} 0 & \text{if the profile } \mathbf{g}_i^+ \text{ properly identifies tumor tissues} \\ 1 & \text{otherwise} \end{cases}, \tag{2}$$

indicating if the expression profiles of i correctly characterize the state specified by the class value of the corresponding tissues. If the genes discriminant function takes the form $\mathbf{w}'\mathbf{g} - b = 0$, where $\mathbf{w}$ defines the orientation of the hyperplane in the s-dimensional space $\mathbb{R}^s$ and b its offset from the origin, the following mixed-integer optimization problem can be formulated

$$\min \sum_{i=1}^{f} \left(z_i^- + z_i^+ \right) \tag{GS}$$

$$\text{s.t.} \quad \mathbf{w}'\mathbf{g}_i^+ - b \geq -Q z_i^+ \qquad i \in \mathcal{F}, \tag{3}$$

$$\mathbf{w}'\mathbf{g}_i^- - b < Q z_i^- \qquad i \in \mathcal{F}, \tag{4}$$

$$\mathbf{z}^-, \mathbf{z}^+ \text{ binaries}, \quad \mathbf{w}, b \text{ free},$$

where Q is a sufficiently large constant scalar, and constraints (3) and (4) set the values of the binary variables z_i^+ and $z_i^-, i \in \mathcal{F}$.

From the solution of problem (GS), obtained by a truncated branch-and-bound procedure, it is possible to find a set of genes useful for discriminating between normal and tumor tissues. In particular, for each gene $i \in \mathcal{F}$ the following measure, termed *classification score*, is computed

$$CS_i = \delta_i^- + \delta_i^+, \tag{5}$$

where δ_i^- and δ_i^+ represent the Euclidean distances of patterns $\mathbf{g_i^-}$ and $\mathbf{g_i^+}$ from the separating hyperplane. Since high classification scores correspond to genes whose expression profiles appear to be significantly different for sample tissues of opposite classes (Fig. 3), a ranking of the genes is generated by sorting them in descending order with respect to their classification score. In this way, the Δ most informative genes are situated in the first Δ positions from the top of the ranking, where Δ is a parameter regulating the number of features to be selected. By joining the sets of top-ranked genes for the datasets $G_t, t = 1, 2, \ldots, r$, the final set of features to use for cancer classification is derived. Clearly, different sets may be obtained by varying the value of the parameter Δ.

Notice that the use of optimization allows to perform a multivariate feature selection analysis since, within each dataset G_t, the contributions of the genes are considered simultaneously. Furthermore, the non-zero components of the separating hyperplanes identify the sample tissues responsible for the optimal

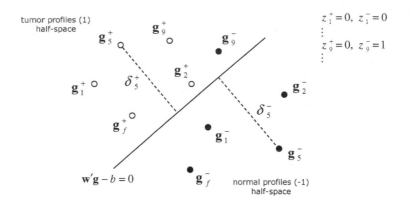

Fig. 3. Gene profiles classification

classification of the gene expression profiles. Thus, the solution of model (GS) for each dataset G_t provides information about the patients who, playing an active role in the gene selection, may deserve particular attention for cancer classification.

Below is provided a synthetic description of the proposed gene selection method.

Gene selection method:
1. Set the iteration counter $t = 1$, and let $G_1, G_2, \ldots, G_r$ be the transposed datasets obtained by the gene expression matrix.
2. While $t < r + 1$, do the following:
   (a) classify the dataset G_t by solving problem (GS);
   (b) for each gene i in G_t compute the classification score CS_i;
   (c) sort the genes in G_t in descending order with respect to the classification score, and select the first Δ genes from the top of the ranking. Let $\mathcal{B}_t$ be the set of top-ranked genes for G_t, and set $t = t + 1$.
3. Build the set of most relevant genes by joining the sets $\mathcal{B}_t, t = 1, 2, \ldots, r$.

---

3   A Polyhedral Classifier for Microarray Data Classification

Once a set of informative genes has been selected, to perform microarray data classification any supervised learning method can be in principle applied. In this paper, we propose to use a polyhedral classifier recently introduced in (Orsenigo and Vercellis, 2007), which represents an extension of *discrete support vector machines* (Orsenigo and Vercellis, 2003; 2004) for solving binary classification problems. This choice is motivated by two main reasons. First, the polyhedral method is able to classify non-linearly separable datasets by means of a set of linear hyperplanes, without resorting to projections into a higher dimensional feature space, as in traditional support vector machines. This helps

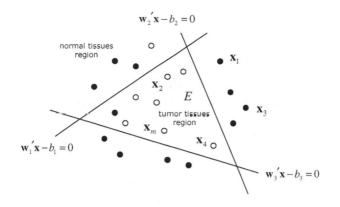

Fig. 4. Polyhedral classification of normal and tumor tissues

preserving the interpretability of the classification rules generated. Furthermore, the polyhedral classifier has proven to learn quite precisely from just a few training examples. Thus, it may be suitable for cancer microarray data classification in which a small number of normal and tumor tissues are usually available.

The polyhedral method generates a classification region E defined by the intersection of the upper half-spaces supported by a set of optimal separating hyperplanes $H_k : \mathbf{w}'_k\mathbf{x} - b_k = 0, k \in \mathcal{L} = \{1, 2, \ldots, L\}$ (Fig. 4). Here L represents an upper bound on the number of hyperplanes the polyhedral method is allowed to generate for the classification of the tissues $\mathbf{x}_i$ in the matrix $\mathcal{S}$. Let $\mathcal{N} = \{1, 2, \ldots, n\}$ be the set of indices of the genes in $\mathcal{S}$. The polyhedral classifier is formulated as the following mixed-integer optimization problem

$$\min \frac{\alpha}{m} \sum_{i=1}^{m} c_i p_i + \beta \sum_{k=1}^{L} \sum_{j=1}^{n} u_{kj} + \gamma \sum_{j=1}^{n} h_j q_j \qquad \text{(PCP)}$$

$$\text{s.t.} \quad \mathbf{w}'_k \mathbf{x}_i - b_k \geq -Q d_{ki} \qquad\qquad i \in \mathcal{M}, k \in \mathcal{L} \qquad (6)$$

$$\mathbf{w}'_k \mathbf{x}_i - b_k \leq (1 - d_{ki})\,Q - \varepsilon \qquad i \in \mathcal{M}, k \in \mathcal{L} \qquad (7)$$

$$e_i \leq \sum_{i=1}^{m} d_{ki} \qquad\qquad i \in \mathcal{M}, k \in \mathcal{L} \qquad (8)$$

$$L e_i \geq \sum_{i=1}^{m} d_{ki} \qquad\qquad i \in \mathcal{M}, k \in \mathcal{L} \qquad (9)$$

$$2v - 1 - y_i(1 - 2e_i) \leq 2p_i \qquad i \in \mathcal{M} \qquad (10)$$

$$2v - 1 - y_i(1 - 2e_i) \geq -2p_i \qquad i \in \mathcal{M} \qquad (11)$$

$$L q_j \geq \sum_{k=1}^{L} f_{kj} \qquad\qquad j \in \mathcal{N} \qquad (12)$$

$$u_{kj} \leq R f_{kj} \qquad\qquad k \in \mathcal{L}, j \in \mathcal{N} \qquad (13)$$

$$w_{kj} \leq u_{kj}, \quad w_{kj} \geq -u_{kj} \qquad k \in \mathcal{L}, j \in \mathcal{N} \qquad (14)$$

$$u_{kj} \le 1 \qquad\qquad\qquad\qquad k \in \mathcal{L}, j \in \mathcal{N} \qquad\qquad (15)$$
$$\mathbf{u} \ge 0, \mathbf{w}_k, \mathbf{b} \qquad\qquad\qquad\qquad k \in \mathcal{L}$$
$$\mathbf{p}, \mathbf{q}, \mathbf{e}, \mathbf{d}, \mathbf{f}, v \text{ binaries.}$$

The objective function of problem (PCP) is composed by the weighted sum of three terms, expressing a trade-off between accuracy and potential of generalization. The first term evaluates the accuracy on $\mathcal{S}$ by means of the *misclassification rate*, defined as the proportion of misclassified sample tissues. This rate is computed by means of the binary variables $p_i, i \in \mathcal{M}$, which take the value 1 if the corresponding tissues are misclassified, and whose values are set by constraints (10) and (11). The second term in the objective function represents the linearized version of the regularization term used in the support vector machines framework (Vapnik, 1995). It is given by the sum of the continuous variables $u_{kj}, k \in \mathcal{L}, j \in \mathcal{N}$, which play a bounding role on the coefficients of $\mathbf{w}_k$ by means of constraints (14) and (15), and allow to formulate a linear optimization problem. Finally, the third term supports the generalization capability of the classifier by minimizing the number of features (genes) defining the classification region E. This number is evaluated by means of two groups of binary variables, f_{kj} and $q_j, k \in \mathcal{L}, j \in \mathcal{N}$. In particular, constraints (13) set the value of f_{kj} to 1 if gene j is active in the definition of $\mathbf{w}_k$, that is if $w_{kj} \ne 0$; constraints (12) force q_j to 1 if gene j is active in the definition of at least one hyperplane, and thereby of the classification region E. The remaining families of binary variables in problem (PCP) are introduced in order to properly count the number of misclassified tissues. In particular, the variables $d_{ki}, k \in \mathcal{L}, i \in \mathcal{M}$, indicate to which half-space supported by H_k the tissues belong, $e_i, i \in \mathcal{M}$, identify the tissues lying inside the classification region E, and v indicates the class value assigned by the model to the tissues lying inside E. Further details on the (PCP) formulation are provided in (Orsenigo and Vercellis, 2007).

Instances of (PCP) of limited size can be solved to optimality by a branch-and-bound procedure. To solve larger instances within reasonable computing times, one may resort to an approximate sequential algorithm by extending to problem (PCP) the heuristic procedure described in (Orsenigo and Vercellis, 2003).

4   Computational Settings and Tests

The usefulness of the proposed method has been evaluated by means of computational tests concerning the classification of three cancer microarray datasets, on which several gene selection and machine learning algorithms have been applied. These datasets, indicated in the sequel as *Ovarian*, *Prostate* and *Colon*, are available at the Kent Ridge Biomedical Data Set Repository (Li and Liu, 2003), and were selected since they are representative of a wide variety of learning tasks. In particular, the *Ovarian* and the *Colon* datasets are interesting for testing the effectiveness of the feature selection procedure, since they contain a huge and a relative low number of genes, respectively. The *Prostate* dataset, instead, is suitable for evaluating the prediction capability of the classification method with

Table 1. Description of the datasets and published classification results in terms of accuracy on the test set (Test acc.) and corresponding number of misclassified sample tissues (Errors)

	Dataset		
Summary	Ovarian	Prostate	Colon
Classes	[Normal(-1),Cancer(1)]	[Normal(-1),Tumor(1)]	[Negative(-1),Positive(1)]
Training set	[50(-1), 50(1)]	[50(-1), 52(1)]	[15(-1), 26(1)]
Test set	[66(-1), 50(1)]	[9(-1), 25(1)]	[7(-1), 14(1)]
No. of genes	15154	12600	2000
Test acc. (%)	$81.0 - 97.4^{(1)}$	$67.6 - 97.1^{(2)}$	$71.4 - 95.2^{(3)(*)}$
Errors (No.)	$22 - 3$	$11 - 1$	$6 - 1$

[1] (Shah and Kusiak, 2007), [2] (Wang et al., 2007), [3] (Li et al., 2001)
[*] The best result 95.2% is obtained on a different test set.

respect to the tumor class, since the training is performed on a balanced dataset and the accuracy is computed on a test set containing a predominant number of tumor tissues. Furthermore, the *Prostate* test set is hard to classify since the sample tissues it contains are obtained from different sources, and appear significantly different with respect to the training set examples. For all the datasets considered the problem is to discriminate between normal and tumor sample tissues. Table 1 provides a summary of the characteristics of the datasets together with the minimum and maximum accuracy values obtained on the corresponding test sets by the methods so far applied.

The first phase of the proposed framework is represented by the selection of the most relevant genes, which was performed according to the procedure described in section 2. In particular, the matrix G obtained for each dataset was randomly partitioned into 10 disjoint subsets $G_1, G_2, \ldots, G_{10}$, so that the generic dataset G_t contained about 3000, 2500 and 400 genes for the *Ovarian*, the *Prostate* and the *Colon* dataset, respectively. Moreover, from each dataset G_t three different sets of top-ranked genes were extracted by letting the parameter Δ varying in the interval $[1, 3]$. In this way, for each classification task three overlapping feature sets were derived, containing the best 10, 20 and 30 genes, respectively.

In order to perform a fair comparison with most of the published results, the accuracy of the polyhedral classifier was assessed by means of holdout estimation (Kohavi, 1995), using the same training and test sets available at the Kent Repository. According to the set of genes used in the training phase, table 2 shows the performances obtained on the test sets by the proposed approach in terms of overall accuracy and *sensitivity*, defined as the percentage of tumor tissues correctly classified. In particular, the values in square brackets in columns 2 and 3 correspond to the total number of misclassified sample tissues and to the number of misclassified tumor examples, respectively. Table 2 includes also two columns which indicate the number of genes used for the classification and

Table 2. Classification results and number of genes

Dataset	Δ	Classification results		No. of genes	L^*
		Overall accuracy % [No. of errors]	Sensitivity % [No. of errors]		
	1	91.4 [10]	90.0 [5]	6	3
Ovarian	2	94.8 [6]	96.0 [2]	7	3
	3	94.8 [6]	96.0 [2]	7	2
	1	88.2 [4]	92.0 [2]	4	3
Prostate	2	94.1 [2]	96.0 [1]	6	2
	3	94.1 [2]	96.0 [1]	7	2
	1	76.2 [5]	85.7 [2]	4	2
Colon	2	85.7 [3]	100 [0]	7	3
	3	80.9 [4]	92.8 [1]	8	3

the number L^* of hyperplanes generated by the polyhedral method for deriving the optimal classification region.

From the results presented in table 2 some empirical conclusions can be drawn. The overall accuracy of the proposed method is competitive to the best performances achieved, being close to the right extreme of the interval of accuracy values provided by competing methods. Notice that, due to the limited size of the test sets, the misclassification of one more tissue leads to a significant drop in the overall accuracy. In particular, the best results are obtained when the classification is performed using the set composed by the 20 top-ranked genes. For *Ovarian* and *Prostate* the largest feature sets provide the highest classification accuracy as well. On the contrary, for the *Colon* dataset the use of the 30 top-ranked genes causes a worsening in the overall performances, probably due to the noise introduced in the classification process by the additional 10 genes included in the feature set.

For what concerns the sensitivity, the results achieved by the proposed approach appear promising. Indeed, the polyhedral classifier is able to perfectly predict the class of most of the tumor tissues in the test sets, at the expense of the misclassification of a low number of normal examples. In the context of cancer classification, the evaluation of sensitivity is fundamental, since the cost of the errors of the first type, represented by the misclassification of tumor tissues, is much greater than the cost associated to second type errors.

Finally, the classification of the datasets is consistently based on a very small number of genes, compared with the original number of features describing the sample tissues. In our computations, these genes can be regarded as the most influential features for tumor differentiation, and lead to simple classification rules which lend themselves to an easier investigation and a less expensive application. From one side, an accurate discrimination between normal and tumor tissues may be achieved by measuring and evaluating the expression profiles of a very small number of marker genes. From the other side, biologists and researchers in bioinformatics may focus their attention on the selected genes in order to study their role in cancer development and evolution.

5 Conclusions and Future Works

We have proposed a novel two-phase framework for gene selection and cancer microarray data classification based on mixed-integer optimization. In the first phase, optimization is used in order to identify small subsets of predictive features, that is genes whose expression profiles are significantly different for normal and tumor sample tissues. In the second phase, tissues of opposite classes are classified by means of a polyhedral method which generates classification regions based on the expression values of the genes selected in the first phase. Computational experiments performed on benchmark datasets indicate that the proposed method exhibits classification results competitive to the best accuracy values so far achieved in the literature, across all the datasets considered. The performances appear promising also in terms of sensitivity. Furthermore, the classification rules generated are consistently based on a very small number of genes and consequently lend themselves to an easier analysis by the domain experts.

Future extensions of this research will concern the development of an approximate algorithm for solving model (GS), alternative to the truncated branch-and-bound procedure, and the investigation of the biological relevance of the genes selected for each cancer classification problem.

References

[Bagui et al., 2003]Bagui, S.C., Bagui, S., Pal, K., Pal, N.R.: Breast cancer detection using ranknearest neighbor classification rules. Pattern Recognition 36, 25–34 (2003)

[Bishop, 1995]Bishop, C.M.: Neural Networks for Pattern Recognition. Oxford Clarendon Press (1995)

[Cho and Won, 2007]Cho, S., Won, H.: Cancer classification using ensemble of neural networks with multiple significant gene subsets. Applied Intelligence 26, 243–250 (2007)

[Cuperlovic-Culf et al., 2005]Cuperlovic-Culf, M., Belacel, N., Ouellette, R.: Determination of tumour marker genes from gene expression data. Drug Discovery Today Target 15, 429–437 (2005)

[Furey et al., 2000]Furey, T.S., Cristianini, N., Duffy, N., Bednarski, M., Schummer, D., Haussler, D.: Support vector machine classification and validation of cancer tissue samples using microarray expression data. Bioinformatics 16, 906–914 (2000)

[Golub et al., 1999]Golub, T.R., Slonim, D.K., Tamayo, P., Huard, C., Gaasenbeek, M., Mesirov, J.P., Coller, H., Loh, M., Downing, L.J.R., Caligiuri, M.A., Bloomeld, C.D., Lander, E.S.: Molecular classification of cancer: class discovery and class prediction by gene expression monitoring. Science 286, 531–537 (1999)

[Kohavi, 1995]Kohavi, R.: A study of cross-validation and bootstrap for accuracy estimation and model selection. In: Proc. of the Fourteenth International Conference on Artificial Intelligence, pp. 1137–1143 (1995)

[Lai et al., 2006]Lai, C., Reinders, M.V., Veer, L.V., Wessels, L.: A comparison of univariate and multivariate gene selection techniques for classification of cancer datasets. BMC Bioinformatics 7, 235 (2006)

[Li and Liu, 2003]Li, J., Liu, H.: Kent Ridge Biomedical Data Set Repository (2003), http://research.i2r.a-star.edu.sg/rp/

[Li et al., 2001]Li, L., Weinberg, C.R., Darden, T.A., Pedersen, L.G.: Gene selection for sample classification based on gene expression data: study of sensitivity to choice of parameters of the GA/KNN method. Bioinformatics 17, 1131–1142 (2001)

[Mallick et al., 2005]Mallick, B.K., Debashis, G., Ghosh, M.: Bayesian classification of tumours by using gene expression data. Journal of the Royal Statistical Society: Series B (Statistical Methodology) 67, 219–234 (2006)

[Orsenigo and Vercellis, 2003]Orsenigo, C., Vercellis, C.: Multivariate classification trees based on minimum features discrete support vector machines. IMA Journal of Management Mathematics 14, 221–234 (2003)

[2004]Orsenigo, C., Vercellis, C.: Discrete support vector decision trees via tabu-search. Journal of Computational Statistics and Data Analysis 47, 311–322 (2004)

[Orsenigo and Vercellis, 2007]Orsenigo, C., Vercellis, C.: Accurately learning from few examples with a polyhedral classifier. Computational Optimization and Applications 38, 235–247 (2007)

[Shah and Kusiak, 2007]Shah, S., Kusiak, A.: Cancer gene search with data-mining and genetic algorithms. Computers in Biology and Medicine 37, 251–261 (2007)

[Vapnik, 1995]Vapnik, V.: The nature of statistical learning theory. Springer, New York (1995)

[Wang et al., 2007]Wang, H.Q., Wong, H.S., Huang, D.S., Shu, J.: Extracting gene regulation information for cancer classification. Pattern Recognition 40, 3379–3392 (2007)

[Wong and Hsu, 2008]Wong, T.T., Hsu, C.H.: Two-stage classification methods for microarray data. Expert Systems with Applications 34, 375–383 (2008)

[Yang et al., 2003]Yang, S., Murali, T.M., Pavlovic, V., Schaffer, M., Kasif, S.: RankGene: identification of diagnostic genes based on expression data. Bioinformatics 19, 1578–1579 (2003)

Detection of Protein Complexes in Protein Interaction Networks Using n-Clubs

Srinivas Pasupuleti

ArsLogica IT Laboratories,
Mezzolombardo (Trento), Italy
srinivas@arslogica.it

Abstract. Protein complexes, identified as functional modules in protein interaction networks, are cellular entities that perform certain biological functions. Revealing these modular structures is significant in understanding how cells function. Protein interaction networks can be constructed by representing nodes as proteins and edges as interactions between proteins. In this paper, we use a graph based distance measure, n-clubs, to detect protein complexes in these interaction networks. The quality of clustering protein interaction networks using n-clubs is comparable to that obtained by best known clustering algorithms applied to various protein networks. Moreover, n-clubs approach is driven by a single parameter n in contrast to other clustering algorithms which have numerous parameters to tune for best results.

1 Introduction

Many complex systems in nature and society can be represented as an intricate web of connections among the units they are made of. These networks when expressed as a graph can reveal significant information not only about the system as a whole, but also allows to identify functionally important parts of the system. Finding structural sub-units (communities) associated with highly interconnected parts is crucial to understand the structural and functional properties of networks. Recently, there has been a growing interest on applying graph theoretical concepts for finding such structural sub-units in biological, social, financial, consumer and co-citation networks [20,3,5]. This work is motivated by current surge in the protein interaction (PPI) data available for different organisms as a result of many high-throughput experimental techniques (e.g. yeast-two-hybrid (Y2H) system and mass spectrometry (MS)). Currently, most proteomics data is available for the model organism *Saccharomyces cerevisiae*, by virtue of the availability of a defined and relatively stable proteome, full genome clone libraries, established molecular biology experimental techniques and an assortment of well designed genomics databases [6,17]. Much larger data sets than this will eventually be available for other well studied model organisms as well as for the human proteome. These complex data sets present a formidable challenge for computational biology to develop automated data mining analysis for knowledge discovery.

E. Marchiori and J.H. Moore (Eds.): EvoBIO 2008, LNCS 4973, pp. 153–164, 2008.
© Springer-Verlag Berlin Heidelberg 2008

One of the important challenges related to proteomics is to understand the relationship between the organization of a network and its function. In particular, it is essential to extract functional modules such as protein complexes or regulatory pathways from global interaction networks. Protein complexes, formed by molecular aggregation of proteins assembled from multiple stable protein interactions, play crucial role in many cellular processes. Identifying protein complexes in the interaction networks can be useful to understand the functions and properties of individual proteins [21]. Predicting molecular complexes from protein interaction data is important also because it provides another level of functional annotation above other guilt-by-association methods. Since sub-units of a molecular complex generally function towards the same biological goal, prediction of an unknown protein as part of a complex also allows increased confidence in the annotation of that protein.

In this paper we present a new graph clustering algorithm based on n-clubs to predict protein complexes from protein-protein interaction graphs. An induced subgraph S is referred to a n-club if the maximum distance between any pair of nodes in S is n. The algorithm starts with a deterministic seed node and builds a subgraph considering all nodes at a distance of n from seed node. In the next step, the nodes with least weight (defined in next section) are removed one after another from the subgraph till we reach a n-club. This process is repeated till all the nodes in the graph are covered. The main characteristics of the proposed algorithm is that it is governed by a single parameter n, finds locally dense subgraphs, and is computationally efficient. The quality of protein complex prediction using n-clubs, is comparable with that obtained by any of the best known clustering algorithm till date.

The n-club algorithm is presented in detail in Section 2. In Section 3, various algorithms used for comparison with n-club algorithm are described. The comparison results with different clustering algorithms is given in Section 4. The papers conclusion and different matching statistics used to evelute predicted clusters is presented in Sections 5 and 6.

2   The Proposed Algorithm

The network of interactions between proteins is generally represented as an interaction graph, where nodes represent proteins and edges represent pairwise interactions. It is suggested that protein complexes correspond dense regions (clusters) in protein interaction networks [8]. The density of a cluster is commonly defined as the ratio of intra-cluster edges to the total possible edges, which is given by $2m/(n(n-1))$, where n is the number of proteins in the cluster and m is the number of interactions between them [22].

The density of a connected component can also be viewed in terms of distance between the nodes, i.e., the minimum number of hops taken to reach from one node to another. In such a case, we define the densest component of the graph as the one in which the distance between any two nodes is one. As the minimum distance between the nodes increases, the density decreases. Thus we can say that density of the component is inversely proportional to the minimum distance between the nodes. Based upon this relation between density and minimum distance between the nodes, we propose an algorithm to detect functional modules in protein interaction networks.

The concept of grouping the nodes into a cluster based on the distance between nodes is appealing due to certain peculiar features of the protein interaction networks. Firstly, the protein networks are sparsely connected, incomplete and noisy. The average density of all the subgraphs of each functional category in MIPS database is averaged about 0.0023. Moreover, the average diameter of the subgraphs of all functional categories in MIPS database is approximately 4 [18]. Hence the nodes which are part of protein complexes, though not strongly connected with each other are at smaller distances from each other. In this paper, we explore how effective can be a clustering approach which groups proteins based on the distance between them.

Before defining n-clubs, we introduce some basic graph terminology as follows:

- A graph $G = (V, E)$ denotes an arbitrary undirected graph, where $V = \{1, 2, \ldots, n\}$ is the set of vertices of G, and $E \subseteq (V \times V - 1)/2$ is the set of edges of G.
- A subgraph G' of a graph G is said to be induced if, for any pair of vertices x and y of G', (x, y) is an edge of G' if and only if (x, y) is an edge of G.
- The distance $d_G(u, v)$ between two vertices u and v in graph G is the number of edges in a shortest path connecting them.
- The diameter $\text{diam}(G)$ of a graph G is the greatest distance between any pair of vertices in G.

Hence, we define n-clubs as follows:

Definition 1. *An induced subgraph G' of a graph G is said to form an n-club if the $\text{diam}(G')$ is n or less ($\text{diam}(G' \leq n)$).*

The algorithm to find clusters based on n-clubs is shown in Figure 1. The input to the algorithm is an undirected simple graph G and the value of diameter n in n-clubs to be found. The three main steps of the algorithm are selection of seed node, formation of cluster (n-club) and check for termination.

Seed Selection

Each cluster starts at a deterministic single node which we call the seed node. The highest weighted node in graph G is considered as the seed node (line 4). In case the highest node-weight is zero, the highest degree node is considered as the seed node. The weight of each node is calculated based on the following definition:

Definition 2. *The weight $w_u(G)$ of a node u is given by the number of edges between its neighbors in graph G.*

Cluster Formation

Once the seed node is determined, we construct a subgraph G' from graph G using the method buildSubgraph(G, s, n), where s is the seed node and n is the

---

$N(G)$	set of nodes in graph G
$diam(G)$	diameter of graph G
s	seed node
n	diameter in n-club
$w_u(G)$	weight of node u in graph G

---

1. **function** n-clubs (G, n)
2.   do
3.   **while** $(N(G) \neq \emptyset)$
4. $s \leftarrow$ arg $max\{w_0(G), w_1(G), \ldots, w_{N(G)}(G)\}$
5. $G' \leftarrow$ buildSubgraph (G, s, n)
6. $diam(G') \leftarrow$ calcDiameter (G', n)
7.   **while**$(diam(G') > n)$
8. $s' \leftarrow$ arg $min\{w_0(G'), w_1(G'), \ldots, w_{N(G')}(G')\}$
9. $G' \leftarrow G' - s'$
10. $diam(G') \leftarrow$ calcDiameter (G', n)
11.   print G'
12. $G \leftarrow G - G'$
13. **return**;

Fig. 1. Clustering Algorithm based on n-clubs

maximum distance between any two nodes of n-club (line 5). The subgraph G' from G is constructed starting from the seed node s and including every node in graph G which is at a distance of n or less. This effectively includes every possible node into G' which can be part of the n-club formed with seed node as the reference node. The resulting subgraph G' from the method buildSubgraph (G, s, n) is then pruned to form an n-club by removing the least weighted nodes one by one till its diameter is less than or equal to n (lines $6 - 10$). This ensures that the final n-club we discover from subgraph G' is composed of highly-connected nodes.

After every node removal, the diameter of the subgraph G' is calculated using the method, calcDiameter(G', n). It starts by representing the subgraph G' as a tree, considering the least weighted node of the subgraph G' as root vertex. The neighbors of root vertex in G' become the nodes at level one of the tree, i.e., children of root vertex. The neighbors of all the nodes at level one which are not already part of the tree will be the nodes at level two and so on. The diameter of subgraph G' is equivalent to the number of levels in the constructed tree. The time-complexity to build the entire tree is given by $O(|V| + |E|)$, where $|V|$ is the set of nodes and $|E|$ is the set of edges in subgraph G'.

In our algorithm we only need to check whether diameter of G' is greater than n (line 6) and hence do not always need to build the complete tree. As the number of levels in the tree is equivalent to its diameter, checking for diameter greater than n is equivalent to building the tree till we reach the level equal to $n+1$ or till there are no nodes to be added, whichever is first. If the set of nodes from G' to be added to the tree become empty before reaching the level $n+1$ of the tree, then the diameter of G' is less than or equal to n, otherwise greater than n.

The choice of root vertex also plays a role in the computation of the algorithm. If we choose a highly weighted node in subgraph G' as root vertex, we need to traverse many nodes to reach the level $n+1$. Hence, the best choice is to start building the tree with least weighted node as the root vertex. This node also happens to be least connected in subgraph G'. Neverthless, it is important to note that if the graph has more than one least weighted node, then all those nodes should be tried as root vertex before determining the diameter of the graph. The maximum value of diameter thus obtained is the actual diameter of the graph.

Termination and Output

Once a cluster is identified with diameter less than n, it is removed from the graph G and printed out. The next cluster is then formed in the remaining graph by identifying a new seed node and the process is repeated until no node is left in the graph G.

3   Clustering Algorithms Used for Comparison

We compare our n-club algorithm with a varied set of clustering algorithms used for finding protein complexes. A brief description of each of the algorithm is presented below.

The Markov Cluster algorithm (MCL) [7] simulates a flow on the graph by calculating successive powers of the associated adjacency matrix. At each iteration, an *inflation step* is applied to enhance the contrast between regions of strong or weak flow in the graph. The process then converges towards a partition of the graph, with a set of high-flow regions (clusters) separated by boundaries with no flow. The value of the *inflation parameter* strongly influences the number of clusters.

Molecular Complex Detection (MCODE) [1], detects densely connected regions. First it assigns a weight to each vertex, corresponding to its local neighborhood density. Then, starting from the top weighted vertex (seed vertex), it recursively moves outward, including in the cluster vertices whose weight is above a given threshold. This threshold corresponds to a user-defined percentage of the weight of the seed vertex. However, since the highly weighted vertices may not be highly connected to each other, the algorithm does not guarantee that the discovered regions are dense.

Restricted Neighborhood Search Clustering (RNSC) [19] , is a cost-based local search algorithm that explores the solution space to minimize a cost function which is calculated according to the number of intra cluster and inter-cluster edges. Starting from an initial random solution, RNSC iteratively moves a vertex from one cluster to another if this move reduces the general cost. When a (user-specified) number of moves has been reached without decreasing the cost function, the algorithm terminates. However, the algorithm results are highly dependent on the quality of initial clustering which is random or user-defined and moreover the algorithm is governed by a number of parameters whose tuning is an overhead to the user.

Super Paramagnetic Clustering (SPC) [2] uses an analogy to the physical properties of an inhomogeneous ferromagnetic model to find tightly connected clusters in a large graph. At first, SPC associates a *spin* with each node of the graph. Spins belonging to highly connected region fluctuate in a correlated fashion and nodes with correlated spins are placed in same cluster. When the temperature increases, the system becomes less stable and the clusters become smaller. The SPC method is best at detecting high density clusters with relatively few links with the outside world, but is not very effective for sparsely connected graphs such as those formed from high-throughput data sets.

4   Experimental Results

The most commonly used high throughput analysis methods for detecting protein interactions are Yeast two-Hybrid (Y2H) system and Mass Spectrometry(MS). Protein interaction network can be constructed by representing nodes as proteins and edges as interaction between proteins. In this work we use six data high-throughput data sets collected from BIO-GRID [9,4]. The data sets represent the network of protein interactions in the yeast *Saccharomyces cerevisiae*. Two of these data sets consists of pairs of interacting protein detected by the two-hybrid technique published respectively by Uetz *et al.* [16] and Ito *et al.* [15]. The four other data sets contain protein complexes characterized by mass spectrometry, published respectively by Gavin *et al.* [10,11], Ho *et al.* [13], and Krogan *et al.* [14]. The data sets are named after the first authors who discovered these data sets and in some cases, the name is superseded by year if there is more than one data set discovered by single author.

The quality of clustering results is evaluated using the statistics, Sensitivity (Sn), Positive Predicted Value (PPV) and Accuracy (Acc). These are calculated by comparing the clusters with every annotated complex. To assess our predictions, we use the curated protein complexes in MIPS [12]. The *complex-wise sensitivity* (Sn) represents the coverage of a complex by its best-matching cluster, given by the maximal fraction of proteins in the complex found in a common cluster. Reciprocally, the *cluster-wise Positive Predicted Value* (PPV) measures how well a given cluster predicts its best matching complex.

To estimate the overall correspondence between a clustering result (a set of clusters) and the collection of annotated complexes, we compute the weighted means of all PPV values (averaged over all clusters) and Sn values (averaged over all complexes). The accuracy (Acc) of the algorithm prediction is given by the geometric mean of the averaged Sn and PPV values (see section *Methods* for a detailed description of the matching statistics).

We ran the n-club algorithm on the six data sets collected for different values of n ranging from 1 to 20. The clusters obtained from these high-throughput networks were compared with the complexes annotated in the MIPS database to compute Sensitivity (Sn), Positive predicted value (PPV) and Accuracy (Acc). It is not straight forward to interpret the results of these measures, in particular "positive predictive value". The reference set of MIPS complexes filtered to discard any high-throughput result, is by no means exhaustive since the complexes detected by previous studies represent only a fraction of all existing complexes. High-throughput methods are thus expected to yield many complexes that have not previously been characterized by other methods.Thus, interactions detected by high-throughput methods that are not annotated in MIPS cannot be considered "false positives". Hence, PPV values should be interpreted as an indication of the fraction of high-throughput results which are also detected by other methods and have been annotated in the MIPS so far. In contrast, the sensitivity is likely to yield more direct relevant information, by indicating the fraction of annotated complexes recovered in the clusters obtained from high-throughput data.

Figure 2 shows the variation of Sn, PPV and Acc with respect to diameter n in n-clubs. These plots present an overview of the behavior of the n-club clustering algorithm for different values of n in n-clubs an also help us to find the optimal value of n to detect high quality complexes in protein interaction networks.

The variation of Sensitivity (Sn) and PPV with respect to diameter n in n-club is shown in Figures 2(a) - 2(b). These measures provide complementary and somewhat contradictory information: when the number of clusters decreases, the Sn increases and, in the trivial case where all proteins are grouped in a single

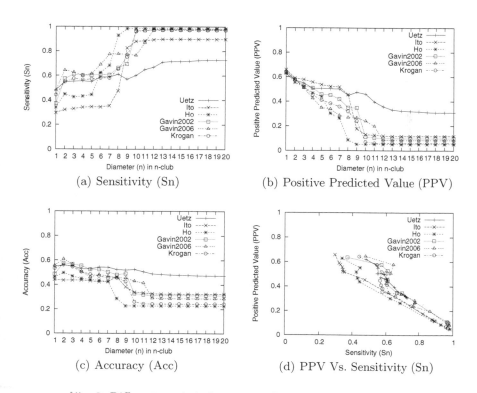

(a) Sensitivity (Sn)

(b) Positive Predicted Value (PPV)

(c) Accuracy (Acc)

(d) PPV Vs. Sensitivity (Sn)

Fig. 2. Different statistical measures for n-club based clustering

cluster, the calculated Sn reaches 1. Reciprocally, the PPV increases with the number of clusters, reaching 1 in the trivial case where each protein is assigned to one separate cluster.

The number of clusters found using n-clubs decreases with increase in diameter, as larger n makes large number of nodes to be grouped into few clusters. The maximum number of clusters is found with value of n equal to one, which also happen to be the densest component of the network. Thus, Sn is the minimum and PPV is the maximum for n equal to one. The statistics Sn and PPV reflect popularly known "*precision-recall*" measures. With increase in n, Sn increases and PPV decreases, as the number of clusters found decreases. For values of n greater than 10, maximum number of nodes of the network are grouped into few clusters, thus making Sn reach its maximum and PPV reach its minimum.

Figure 2(c) shows the variation of Accuracy(Acc) with diameter n in n-clubs. Accuracy is given by the geometric mean of Sn and PPV. The graph shows that high accuracy values are obtained for a diameter of either 2 or 3, which is a reasonable relaxation of clique definition. The accuracy decreases with increase in the value of n due to the higher decrease in PPV compared to increase in Sn. The accuracy for $Uetz$ data set remains stable as it has the smallest density (0.018) of all the networks and hence the increase in n has little effect on Sn and PPV.

Figure 2(d) is a typical "*precision-recall*" plot, where Sensitivity (Sn) reflects the recall and PPV reflects precision. The figure shows the typical behavior of a precision-recall plot with PPV (precision) inversely proportional to Sn (recall). The values of Sn and PPV are obtained for different values of n in n-clubs over all the data sets. The precision-recall plots help in identifying the parameter values for best performance of the clustering algorithm. Ideally, the performance of an algorithm is optimal when both, Sn and PPV, are high and this point is given by the break-even point in the precision-recall plot. Break-even point is defined as the point where precision equals recall. Some precision recall graphs do not have a break-even point as defined above. An alternative definition for break-even point can be the point with the smallest difference between precision and recall among all those that have larger precision than recall. This refers to the points close to the intersection of straight line drawn onto the precision-recall plot passing through origin with a slope equal to one. It has been found from our analysis that for all the protein data sets, the break-even point is obtained for values of n equal to either 2 or 3.

The results thus obtained by fixing the diameter of n-clubs at 2, are compared with four different clustering algorithms discussed in Section 3. The performance statistics $(Sn, PPV$ and $Acc)$ for MCL, MCODE, RNSC and SPC algorithms are taken from Sylvian [4], who did experiments on high-throughput data sets by optimizing the parameters for best performance of each of the algorithm. The comparison results along with number of nodes and edges for each data set are given in Table 1. The values of Sn, PPV and Acc are converted to a percentage (x 100), for better readability.

Table 1. Statistical comparison of different clustering algorithms on high-throughput data sets

Dataset	No. of Nodes	No. of Edges	Statistics	N-clubs	MCL	MCODE	RNSC	SPC
Uetz	926	865	Sn	54.65	57.3	84.3	49.4	65.5
			PPV	57.26	53.8	25.5	59.6	38.0
			Acc	55.94	55.52	46.36	54.26	49.88
Ito	2937	4038	Sn	32.15	34.9	66.9	31.4	73.2
			PPV	59.0	42.7	8.2	63.6	24.3
			Acc	43.55	38.6	23.42	44.68	42.17
Ho	1564	3600	Sn	44.69	50.6	81.2	37.0	90.1
			PPV	55.3	47.1	12.9	61.5	10.4
			Acc	49.71	48.81	32.36	47.7	30.61
Gavin2002	1352	3210	Sn	57.36	74.1	67.0	52.1	91.8
			PPV	57.71	57.0	20.4	62.0	18.1
			Acc	57.53	65.0	37.0	56.83	40.76
Gavin2006	1430	6531	Sn	64.26	75.7	58.3	60.8	79.8
			PPV	57.85	54.3	20.6	63.3	37.0
			Acc	60.97	64.11	34.65	62.03	54.33
Krogan	2675	7088	Sn	54.82	62.8	56.3	53.1	82.6
			PPV	57.58	56.2	21.9	63.3	25.4
			Acc	56.18	59.4	35.11	58.0	45.8

The results show that n-club approach achieves high accuracy for the data sets *Utez* and *Ho* and is second best for the data sets *Ito* and *Gavin*2002. The difference between Sn and PPV obtained by n-club approach is small compared to other algorithms, suggesting that n-club algorithm achieves a good balance between the two quality measures.

We also observe that the quality measures obtained by n-club algorithm are approximately close to that obtained by RNSC algorithm, however n-clubs algorithm gains over RNSC with respect to the number of governing parameters. The number of parameters in n-clubs algorithm is one as compared to five in RNSC algorithm [4,19].

5   Conclusions

This work presents a new graph clustering algorithm based on n-clubs to detect functional modules in protein interaction networks. Since protein complexes in interaction networks, on an average, have less diameter and are sparsely connected [18], we find that proteins which are at a smaller distance from each other usually cluster together representing molecular biological functional units. Therefore, it is advantageous to group proteins into clusters based on distance between them, i.e., using n-clubs, to accurately predict protein complexes. The effectiveness of n-club algorithm is shown by comparsion results with four distinct clustering algorithms. Besides yielding high accuracy of protein complex predicition, n-club algorithm also achieves good trade-off between the contrasting quality measures Sn and PPV, compared to other clustering algorithms. Moreover, the proposed algorithm depends on a single parameter n.

A better comparison of different clustering algorithms can be done by plotting Sn-PPV ($precision-recall$) curve for each of them. This requires the algorithms to be tested on all the combination of their parameters. RNSC algorithm is guided by more than five parameters and MCODE has four distinct governing parameters [4], thus leading huge permutation of parameters. The on-going work involves comparing the performance of different clustering algorithms over all the parameter combinations to find the robustness of each of them.

6   Methods

Sensitivity, Positive Predicted Value and Accuracy are classically used to measure the correspondence between the result of a classification and a reference. However, these concepts can be adapted for measuring the match between a set of protein complexes and a clustering result.

Sensitivity
Considering the annotated complexes from MIPS database [12] as our reference classification, we define *sensitivity* as the fraction of proteins of complex i which are found in cluster j.

$$Sn_{i,j} = T_{i,j}/N_i \tag{1}$$

where N_i is the number of proteins in complex i and $T_{i,j}$ is the number of proteins of cluster j which are part of complex i. The *complex-wise sensitivity* Sn_{co_i} is calculated as the maximal fraction of proteins of complex i assigned to the same cluster. Sn_{co_i} reflects the coverage of complex i by its best-matching cluster.

$$Sn_{co_i} = max_{j=1}^{m} Sn_{i,j} \tag{2}$$

To characterize the general sensitivity of a clustering result, we compute a *clustering-wise sensitivity* as the weighted average of Sn_{co_i} over all complexes.

$$Sn = \frac{\sum_{i=1}^{n} N_i Sn_{co_i}}{\sum_{i=1}^{n} N_i} \tag{3}$$

In simple terms, sensitivity as a whole can be referred as "**recall**", used most often in retrieval systems.

Positive Predictive Value
The positive predicted value is the proportion of members of cluster j which belong to complex i, relative to the total number of members of this cluster assigned to all complexes

$$PPVi,j = T_{i,j} / \sum_{i=1}^{n} T_{i,j} \tag{4}$$

The summation in the denominator, given by T_j, is the total number of proteins belonging to all possible complexes. A protein is counted more than one if it is part of more than one complex. Hence, in some cases it may differ from the cluster size, because some proteins can belong to several complexes.

The *cluster-wise positive predicted value*, PPV_{cl_j}, is calculated as the maximal fraction of proteins of cluster j found in the same annotated complex. PPV_{cl_j} reflects the reliability with which cluster j predicts that a protein belongs to its best-matching complex.

$$PPV_{cl_j} = max_{i=1}^{n} PPV_{i,j} \tag{5}$$

To characterize the general PPV of a clustering result as a whole, we compute a *clustering-wise PPV* as the weighted average of PPV_{cl_j} over all clusters.

$$PPV = \frac{\sum_{j=1}^{m} T_j PPV_{cl_j}}{\sum_{j=1}^{m} T_j} \tag{6}$$

PPV can be seen equivalent to "**precision**" used most often in retrieval systems.

Accuracy

The *geometric accuracy* (*Acc*) indicates the tradeoff between sensitivity and predicted value. It is obtained by computing the geometrical mean of the Sn and the PPV.

$$Acc = \sqrt{Sn.PPV} \tag{7}$$

The advantage of taking the geometric rather than arithmetic mean is that it yields a low score when either the Sn or the PPV metric is low. High accuracy values thus require a high performance for both criteria.

Acknowledgement. I am grateful to Orkun S. Soyer, Researcher at Microsoft Research-University of Trento Centre for Computational and Sytsems Biology, for useful discussions and feedback.

References

1. Bader, G., Hogue, C.: An automated method for finding molecular complexes in large protein interaction networks. BMC Bioinformatics 4(1), 2 (2003)
2. Blatt, M., Wiseman, S., Domany, E.: Superparamagnetic clustering of data. Phys Rev Lett 76, 3251–3254 (1996)
3. Boginski, V., Butenko, S., Pardalos, P.M.: Mining market data: A network approach. Comput. Oper. Res. 33(11), 3171–3184 (2006)
4. Brohee, S., van Helden, J.: Evaluation of clustering algorithms for protein-protein interaction networks. BMC Bioinformatics 7(488) (2006)

5. Butenko, S., Wilhelm, W.: Clique-detection models in computational biochemistry and genomics. European Journal of Operational Research 173, 1–17 (2006)
6. Deane, C.M., Salwinski, L., Xenarios, I., Eisenberg, D.: Protein interactions: two methods for assessment of the reliability of high throughput observations. Molecular and Cellular Proteomics 1, 349–356 (2002)
7. Enright, A.J., Van Dongen, S., Ouzounis, C.A.: An efficient algorithm for large-scale detection of protein families. Nucleic Acids Res 30, 1575–1584 (2002)
8. Tong, A.H.Y., et al.: A combined experimental and computational strategy to define protein interaction networks for peptide recognition modules. Science 295(5553), 321–324 (2002)
9. Stark, C., et al.: Biogrid: a general repository for interaction datasets. Nucl. Acids Res. 539, D535–539 (2006)
10. Anne-Claude, G., et al.: Functional organization of the yeast proteome by systematic analysis of protein complexes. Nature 415, 141–147 (2002)
11. Anne-Claude, G., et al.: Proteome survey reveals modularity of the yeast cell machinery. Nature 440, 631–636 (2006)
12. Mewes, H.W., Frishman, D., et al.: MIPS: A database for genomes and protein sequences. Nucliec Acids Research 28, 74–78 (2000)
13. Ho, Y., Gruhler, A., et al.: Systematic identification of protein complexes in saccharomyces cerevisiae by mass spectrometry. Nature 415, 180–183 (2002)
14. Krogan, N.J., et al.: Global landscape of protein complexes in the yeast saccharomyces cerevisiae. Nature 440, 637–643 (2006)
15. Ito, T., et al.: A comprehensive two-hybrid analysis to explore the yeast protein interactome. Proc. National Academy of Sciences, USA 98(8), 4569–4574 (2001)
16. Uetz, et al.: A comprehensive analysis of protein-protein interactions in saccharomyces cerevisiae. Nature 203, 623–627 (2000)
17. Hishgak, H., Nakai, K., Ono, T., Tanigami, T., Takagi, T.: Assessment of prediction accuracy of protein function from protein-protein interaction data. Yeast 18, 523–531 (2001)
18. Hwang, W., Cho, Y.-R., Zhang, A., Ramanathan, M.: A novel functional module detection algorithm for protein-protein interaction networks. Algorithms for Molecular Biology 1(24) (2006)
19. King, A.D., Przulj, N., Jurisica, I.: Protein complex prediction via cost-based clustering. Bioinformatics 20(17), 3013–3020 (2004)
20. Palla, G., Derenyi, I., Farkas, I., Vicsek, T.: Uncovering the overlapping community structure of complex networks in nature and society. Nature 435, 814–818 (2005)
21. Pereira-Leal, J.B., Enright, A.J., Ouzounis, C.A.: Detection of functional modules from protein interaction networks. Proteins 54, 49–57 (2004)
22. Spirin, V., Mirny, L.A.: Protein complexes and functional modules in molecular networks. PNAS 100(21), 12123–12128 (2003)

Learning Gaussian Graphical Models of Gene Networks with False Discovery Rate Control

Jose M. Peña

IFM, Linköping University, SE-58183 Linköping, Sweden
jmp@ifm.liu.se

Abstract. In many cases what matters is not whether a false discovery is made or not but the expected proportion of false discoveries among all the discoveries made, i.e. the so-called false discovery rate (FDR). We present an algorithm aiming at controlling the FDR of edges when learning Gaussian graphical models (GGMs). The algorithm is particularly suitable when dealing with more nodes than samples, e.g. when learning GGMs of gene networks from gene expression data. We illustrate this on the Rosetta compendium [8].

1 Introduction

Some models that have received increasing attention from the bioinformatics community as a means to gain insight into gene networks are Gaussian graphical models (GGMs) and variations thereof [4,6,9,14,25,26,29,31,32]. The GGM of a gene network represents the network as a Gaussian distribution over a set of random variables, each of them representing (the expression level of) a gene in the network. Learning the GGM reduces to learning the independence structure of the Gaussian distribution. This structure is represented as an undirected graph such that if two sets of nodes are separated by a third set of nodes in the graph, then (the expression level of) the corresponding sets of genes are independent given (the expression level of) the third set of genes in the gene network. Gene dependencies can also be read off a GGM [18]. A further advantage of GGMs is that there already exists a wealth of algorithms for learning GGMs from data. However, not all of them are applicable when the database contains fewer samples than nodes (i.e. $n < q$),[1] which is the case in most gene expression databases. Of the algorithms that are applicable when $n < q$, only the one proposed in [25] aims at controlling the false discovery rate (FDR), i.e. the expected proportion of falsely discovered edges among all the edges discovered. However, the correctness of this algorithm is neither proven nor fully supported by the experiments reported, e.g. see the results for sample size 50 in Figure 6 in [25].

In this paper, we present a modification of the incremental association Markov boundary (IAMB) algorithm [19,28] aiming at controlling the FDR. Although

---

[1] We denote the number of nodes (i.e. genes) by q though it is customary to use p for this purpose. We reserve p to denote a probability distribution.

E. Marchiori and J.H. Moore (Eds.): EvoBIO 2008, LNCS 4973, pp. 165–176, 2008.
© Springer-Verlag Berlin Heidelberg 2008

we have not yet succeeded in proving that the new algorithm controls the FDR, the experiments reported in this paper support this conjecture. Furthermore, the new algorithm is particulary suitable for those domains where $n < q$, which makes it attractive for learning GGMs of gene networks from gene expression data. We show that the new algorithm is indeed able to provide biologically insightful models by running it on the Rosetta compendium [8].

2 Preliminaries

The definitions and results in the following two paragraphs are taken from [11,15,27,30]. We use the juxtaposition $\mathbf{XY}$ to denote $\mathbf{X} \cup \mathbf{Y}$, and X to denote the singleton $\{X\}$. Let $\mathbf{U}$ denote a set of q random variables. Unless otherwise stated, all the probability distributions and graphs in this paper are defined over $\mathbf{U}$. Let $\mathbf{X}$, $\mathbf{Y}$, $\mathbf{Z}$ and $\mathbf{W}$ denote four mutually disjoint subsets of $\mathbf{U}$. We represent that $\mathbf{X}$ is independent of $\mathbf{Y}$ given $\mathbf{Z}$ in a probability distribution p by $\mathbf{X} \perp \mathbf{Y}|\mathbf{Z}$, whereas we represent that $\mathbf{X}$ is dependent of $\mathbf{Y}$ given $\mathbf{Z}$ in p by $\mathbf{X} \not\perp \mathbf{Y}|\mathbf{Z}$. Any probability distribution satisfies the following four properties: Symmetry $\mathbf{X} \perp \mathbf{Y}|\mathbf{Z} \Rightarrow \mathbf{Y} \perp \mathbf{X}|\mathbf{Z}$, decomposition $\mathbf{X} \perp \mathbf{YW}|\mathbf{Z} \Rightarrow \mathbf{X} \perp \mathbf{Y}|\mathbf{Z}$, weak union $\mathbf{X} \perp \mathbf{YW}|\mathbf{Z} \Rightarrow \mathbf{X} \perp \mathbf{Y}|\mathbf{ZW}$, and contraction $\mathbf{X} \perp \mathbf{Y}|\mathbf{ZW} \wedge \mathbf{X} \perp \mathbf{W}|\mathbf{Z} \Rightarrow \mathbf{X} \perp \mathbf{YW}|\mathbf{Z}$. Any strictly positive probability distribution also satisfies intersection $\mathbf{X} \perp \mathbf{Y}|\mathbf{ZW} \wedge \mathbf{X} \perp \mathbf{W}|\mathbf{ZY} \Rightarrow \mathbf{X} \perp \mathbf{YW}|\mathbf{Z}$. Any Gaussian distribution also satisfies composition $\mathbf{X} \perp \mathbf{Y}|\mathbf{Z} \wedge \mathbf{X} \perp \mathbf{W}|\mathbf{Z} \Rightarrow \mathbf{X} \perp \mathbf{YW}|\mathbf{Z}$.

Let $sep(\mathbf{X}, \mathbf{Y}|\mathbf{Z})$ denote that $\mathbf{X}$ is separated from $\mathbf{Y}$ given $\mathbf{Z}$ in an undirected graph (UG) G, i.e. every path in G between $\mathbf{X}$ and $\mathbf{Y}$ contains some $Z \in \mathbf{Z}$. G is an undirected independence map of a probability distribution p when $\mathbf{X} \perp \mathbf{Y}|\mathbf{Z}$ if $sep(\mathbf{X}, \mathbf{Y}|\mathbf{Z})$. G is a minimal undirected independence (MUI) map of p when removing any edge from G makes it cease to be an independence map of p. MUI maps are also called Markov networks. Furthermore, p is faithful to G when $\mathbf{X} \perp \mathbf{Y}|\mathbf{Z}$ iff $sep(\mathbf{X}, \mathbf{Y}|\mathbf{Z})$. A Markov boundary of $X \in \mathbf{U}$ in p is any subset $MB(X)$ of $\mathbf{U} \setminus X$ such that (i) $X \perp \mathbf{U} \setminus MB(X) \setminus X | MB(X)$, and (ii) no proper subset of $MB(X)$ satisfies (i). If p satisfies the intersection property, then (i) $MB(X)$ is unique for each $X \in \mathbf{U}$, (ii) the MUI map G of p is unique, and (iii) two nodes X and Y are adjacent in G iff $X \in MB(Y)$ iff $Y \in MB(X)$ iff $X \not\perp Y | \mathbf{U} \setminus (XY)$. The MUI map of a Gaussian distribution p is usually called the Gaussian graphical model (GGM) of p. GGMs are also called covariance selection models. In a Gaussian distribution $Normal(\mu, \Sigma)$, $X \perp Y|\mathbf{Z}$ iff $\rho_{XY|\mathbf{Z}} = 0$, where $\rho_{XY|\mathbf{Z}} = \dfrac{-((\Sigma_{XY\mathbf{Z}})^{-1})_{XY}}{\sqrt{((\Sigma_{XY\mathbf{Z}})^{-1})_{XX}((\Sigma_{XY\mathbf{Z}})^{-1})_{YY}}}$ is the population partial correlation between X and Y given $\mathbf{Z}$.

Assume that a sample of size n from a Gaussian distribution $Normal(\mu, \Sigma)$ is available. Let $r_{XY|\mathbf{Z}}$ denote the sample partial correlation between X and Y given $\mathbf{Z}$, which is calculated as $\rho_{XY|\mathbf{Z}}$ but replacing $\Sigma_{XY\mathbf{Z}}$ by its maximum likelihood estimate based on the sample. Under the null hypothesis that $\rho_{XY|\mathbf{Z}} = 0$, the test statistic $\frac{1}{2} \log \frac{1+r_{XY|\mathbf{Z}}}{1-r_{XY|\mathbf{Z}}}$ has an asymptotic $Normal(0, \frac{1}{\sqrt{n-3-|\mathbf{Z}|}})$ distribution [2]. Moreover, this hypothesis test is consistent [10]. We call this test

Fisher's z-test. Under the null hypothesis that $\rho_{XY|\mathbf{Z}} = 0$, the test statistic $\frac{\sqrt{n-2-|\mathbf{Z}|}\cdot r_{XY|\mathbf{Z}}}{\sqrt{1-r^2_{XY|\mathbf{Z}}}}$ has an exact Student's t distribution with $n - 2 - |\mathbf{Z}|$ degrees of freedom [2]. We call this test Fisher's t-test. Note that Fisher's z-test and t-test are applicable only when $n > |XY\mathbf{Z}|$: These tests require $r_{XY|\mathbf{Z}}$ which in turn requires the maximum likelihood estimate of $\Sigma_{XY\mathbf{Z}}$, and this exists iff $n > |XY\mathbf{Z}|$ [11].

In many problems what matters is not whether a false discovery is made or not but the expected proportion of false discoveries among all the discoveries made. False discovery rate (FDR) control aims at controlling this proportion. Moreover, FDR control tends to have more power than familywise error rate control, which aims at controlling the probability of making some false discovery [3]. Consider testing m null hypotheses $H_0^1, \ldots, H_0^m$. The FDR is formally defined as the expected proportion of true null hypotheses among the null hypotheses rejected, i.e. $FDR = E[|F|/|D|]$ where $|D|$ is the number of null hypotheses rejected (i.e. discoveries) and $|F|$ is the number of true null hypotheses rejected (i.e. false discoveries). Let $p_1, \ldots, p_m$ denote p-values corresponding to $H_0^1, \ldots, H_0^m$. Moreover, let $p_{(i)}$ denote the i-th smallest p-value and $H_0^{(i)}$ its corresponding hypothesis. The following procedure controls the FDR at level α (i.e. $FDR \leq \alpha$) [3]: Reject $H_0^{(1)}, \ldots, H_0^{(j)}$ where j is the largest i for which $p_{(i)} \cdot \frac{m}{i} \cdot \sum_{k=1}^{m} \frac{1}{k} \leq \alpha$. We call this procedure BY.

3   Learning GGMs

In this section, we present three algorithms for learning GGMs from data. The third one is the main contribution of this paper, as it aims at learning GGMs with FDR control when $n < q$. Hereinafter, we assume that the GGM to learn is sparse, i.e. it contains only a small fraction of all the $q(q-1)/2$ possible edges. This assumption is widely accepted in bioinformatics for the GGM of a gene network.

3.1   EE Algorithm

One of the simplest algorithms for learning the GGM G of a Gaussian distribution p consists in making use of the fact that an edge $X - Y$ is in G iff $X \not\perp Y | \mathbf{U} \setminus (XY)$. We call this algorithm edge exclusion (EE), as the algorithm can be seen as starting from the complete graph and, then, excluding from it all the edges $X - Y$ for which $X \perp Y | \mathbf{U} \setminus (XY)$. Since EE performs a finite number of hypothesis tests, EE is consistent when the hypothesis tests are so. Recall from Section 2 that consistent hypothesis tests exist. Note that EE with Fisher's z-test or t-test is applicable only when $n > q$, since these tests are applicable only in this case (recall Section 2).

Since EE can be seen as performing simultaneous hypothesis tests, BY can be embedded in EE to control the FDR. Note that Fisher's z-test relies on the asymptotic probability distribution of the test statistic and, thus, may not return

Table 1. IAMB(X) and IAMBFDR(X)

IAMB(X)	IAMBFDR(X)
1 $MB = \emptyset$	1 $MB = \emptyset$
2 **for** i in $1..q-1$ **do**	2 **for** i in $1..q-1$ **do**
3 $p_i = pvalue(X \perp Y_i \mid MB \setminus Y_i)$	3 $p_i = pvalue(X \perp Y_i \mid MB \setminus Y_i)$
4 **for** i in $q-1..1$ **do**	4 **for** i in $q-1..1$ **do**
5 **if** $Y_{(i)} \in MB$ **then**	5 **if** $Y_{(i)} \in MB$ **then**
6 **if** $p_{(i)} > \alpha$ **then**	6 **if** $p_{(i)} \cdot \frac{q-1}{i} \cdot \sum_{k=1}^{q-1} \frac{1}{k} > \alpha$ **then**
7 $MB = MB \setminus Y_{(i)}$	7 $MB = MB \setminus Y_{(i)}$
8 go to line 2	8 go to line 2
9 **for** i in $1..q-1$ **do**	9 **for** i in $1..q-1$ **do**
10 **if** $Y_{(i)} \notin MB$ **then**	10 **if** $Y_{(i)} \notin MB$ **then**
11 **if** $p_{(i)} \le \alpha$ **then**	11 **if** $p_{(i)} \cdot \frac{q-1}{i} \cdot \sum_{k=1}^{q-1} \frac{1}{k} \le \alpha$ **then**
12 $MB = MB \cup Y_{(i)}$	12 $MB = MB \cup Y_{(i)}$
13 go to line 2	13 go to line 2
14 **return** MB	14 **return** MB

p-values but approximate p-values. This may cause that the FDR is controlled only approximately. Fisher's t-test, on the other hand, returns p-values and, thus, should be preferred in practice.

3.2   IAMB Algorithm

EE is based on the characterization of the GGM of a Gaussian distribution p as the UG G where an edge $X - Y$ is in G iff $X \not\perp Y \mid \mathbf{U} \setminus (XY)$. As a consequence, we have seen above that EE is applicable only when $n > q$. We now describe an algorithm that can be applied when $n < q$ under the sparsity assumption. The algorithm is based on the characterization in which an edge $X - Y$ is in G iff $Y \in MB(X)$. Therefore, we first introduce in Table 1 an algorithm for learning MBs that we call IAMB(X), because it is a modification of the incremental association Markov boundary algorithm studied in [19,28]. IAMB(X) receives the target node X as input and returns an estimate of $MB(X)$ in MB as output. IAMB(X) first computes p-values for the null hypotheses $X \perp Y_i \mid MB \setminus Y_i$ with $Y_i \in \mathbf{U} \setminus X$. In the table, $p_{(i)}$ denotes the i-th smallest p-value and $Y_{(i)}$ the corresponding node. Then, IAMB(X) iterates two steps. The first step aims at removing false discoveries from MB by removing the node with the largest p-value if this is larger than α. The second step is run when the first step cannot remove any node from MB, and it aims at adding true discoveries to MB by adding the node with the smallest p-value if this is smaller than α. Note that after each node removal or addition, the p-values are recomputed. The original IAMB(X) executes step 2 while possible and only then executes step 1. This delay in removing nodes from MB may harm performance as the larger MB gets the less reliable the hypothesis tests tend to be. The modified version proposed here avoids this problem by keeping MB as small as possible at all times. We prove in [19] that the original IAMB(X) is consistent, i.e. its output converges in probability to a MB of X, if the hypothesis tests are consistent. The proof also applies to the modified version presented here. The proof relies on the fact that any Gaussian distribution satisfies the composition property. It is this property what allows IAMB(X) to run forward, i.e. starting with $MB = \emptyset$. Recall from Section 2 that consistent hypothesis tests exist.

IAMB(X) immediately leads to an algorithm for learning the GGM G of p, which we just call IAMB: Run IAMB(X) for each $X \in \mathbf{U}$ and, then, link X and Y in G iff X is in the output of IAMB(Y) or Y is in the output of IAMB(X). Note that, in theory, X is in the output of IAMB(Y) iff Y is in the output of IAMB(X). However, in practice, this may not always be true, particulary when working in high-dimensional domains. That is why IAMB only requires one of the two statements to be true for linking X and Y in G. Obviously, IAMB is consistent under the same assumptions as IAMB(X), namely that the hypothesis tests are consistent.

The advantage of IAMB over EE is that it can be applied when $n < q$, because the largest dimension of the covariance matrix for which the maximum likelihood estimate is computed is not $q \times q$ but $s \times s$, where $s - 2$ is the size of the largest MB at line 3 of IAMB(X). We expect that $s \ll q$ under the sparsity assumption. It goes without saying that there are cases when $n < q$ where IAMB is not applicable either, namely those where $n < s$.

3.3   IAMBFDR Algorithm

Unfortunately, IAMB(X) cannot be seen as performing simultaneous hypothesis tests and, thus, BY cannot be embedded in IAMB(X) to control the FDR. In this section, we present a modification of IAMB(X) aiming at controlling the FDR. The modification is based on redefining $MB(X)$ as the set of nodes such that $Y \in MB(X)$ iff $X \not\perp Y | MB(X) \setminus Y$. We now prove that his redefinition is equivalent to the original definition given in Section 2. If $Y \in MB(X)$, then let us assume to the contrary $X \perp Y | MB(X) \setminus Y$. This together with $X \perp \mathbf{U} \setminus MB(X) \setminus X | MB(X)$ implies $X \perp (\mathbf{U} \setminus MB(X) \setminus X) Y | MB(X) \setminus Y$ by contraction, which contradicts the minimality property of $MB(X)$. On the other hand, if $Y \notin MB(X)$ then $X \perp \mathbf{U} \setminus MB(X) \setminus X | MB(X)$ implies $X \perp Y | MB(X) \setminus Y$ by decomposition.

Specifically, we modify IAMB(X) so that the nodes in the output MB are exactly those whose corresponding null hypotheses are rejected when running BY at level α with respect to the null hypotheses $X \perp Y | MB \setminus Y$. In other words, $Y \in MB$ iff $X \perp Y | MB \setminus Y$ according to BY at level α. To implement this modification, we modify the lines 6 and 11 of IAMB(X) as indicated in Table 1. Therefore, the two steps the modified IAMB(X), which we hereinafter call IAMBFDR(X), iterates are as follows. The first step removes from MB the node with the largest p-value if its corresponding null hypothesis is not rejected by BY at level α. The second step is run when the null hypotheses corresponding to all the nodes in MB are rejected by BY at level α, and it adds to MB the node with the smallest p-value among the nodes whose corresponding null hypotheses are rejected by BY at level α.

Finally, we can replace IAMB(X) by IAMBFDR(X) in IAMB and so obtain an algorithm for learning the GGM G of p. We call this algorithm IAMBFDR. It is easy to see that the proof of consistency of IAMB(X) also applies to IAMBFDR(X) and, thus, IAMBFDR is consistent under the same assumptions as IAMB, namely that the hypothesis tests are consistent. Unfortunately, IAMBFDR does not control the FDR: If the true GGM is the empty graph, then the FDR gets arbitrarily close to 1 as q increases, as any edge discovered by IAMBFDR is a false discovery and the probability that IAMBFDR discovers some edge increases with q. However, if we redefine the FDR of IAMBFDR as the expected FDR of IAMBFDR(X) for $X \in \mathbf{U}$, then IAMBFDR does control the FDR if IAMBFDR(X) controls the FDR: If FDR_X denotes the FDR of IAMBFDR(X), then $E[FDR_X] = \sum_{X \in \mathbf{U}} \frac{1}{q} FDR_X \leq \frac{q}{q} \cdot \alpha$. Although we have not yet succeeded in proving that IAMBFDR(X) controls the FDR, the experiments reported in the next section support the conjecture that IAMBFDR controls the FDR in the latter sense.

4   Evaluation

In this section, we evaluate the performance of EE, IAMB and IAMBFDR on both simulated and gene expression data.

4.1   Simulated Data

We consider databases sampled from random GGMs. Specifically, we consider 100 databases with 50, 100, 500 and 1000 instances sampled from random GGMs with 300 nodes. To produce each of these 400 databases, we do not really sample a random GGM but a random Gaussian network (GN) [7]. The probability distribution so sampled is with probability one faithful to a GGM whose UG is the moral graph of the GN sampled [13]. So, this is a valid procedure for sampling random GGMs. Each GN sampled contains only 1 % of all the possible edges in order to model sparsity. The edges link uniformly drawn pairs of nodes. Each node follows a Gaussian distribution whose mean depends linearly on the value of its parents. For each node, the unconditional mean, the parental linear coefficients and the conditional standard deviation are uniformly drawn from [-3, 3], [-3, 3] and [1, 3], respectively. We do not claim that the databases sampled resemble gene expression databases, apart from some sample sizes and the sparsity of the models sampled. However, they make it possible to compute performance measures such as the power and FDR. This will provide us with some insight into the performance of the algorithms in the evaluation before we turn our attention to gene expression data in the next section.

Table 2 summarizes the results of our experiments with Fisher's t-test and $\alpha = 0.01, 0.05$. Each entry in the table is the average of 100 databases sampled from 100 GGMs randomly generated as indicated above. We do not report standard deviation values because they are very small. For EE, *power* is the fraction of edges in the GGM sampled that are in G, whereas FDR is the fraction of edges in G that are not in the GGM sampled. For IAMB(X) and IAMBFDR(X), *power$_X$* is the fraction of nodes in $MB(X)$ that are in the output MB of IAMB(X) or IAMBFDR(X), $FDR_{X,1}$ is the fraction of nodes in MB that are not in $MB(X)$, and $FDR_{X,2}$ is the fraction of nodes Y in MB such that $X \perp Y | MB \setminus Y$. For

Table 2. Performance of the algorithms on simulated data

n	algorithm	$\alpha = 0.01$						$\alpha = 0.05$					
		sec.	$power$	FDR	$\overline{power}$	$\overline{FDR_1}$	$\overline{FDR_2}$	sec.	$power$	FDR	$\overline{power}$	$\overline{FDR_1}$	$\overline{FDR_2}$
50	IAMB	4	0.49	–	0.45	0.53	0.19	4	–	–	–	–	–
	IAMBFDR	1	0.36	–	0.35	0.05	0.00	1	0.39	–	0.37	0.05	0.00
100	IAMB	4	0.59	–	0.52	0.46	0.19	42	0.65	–	0.57	0.82	0.37
	IAMBFDR	2	0.47	–	0.43	0.04	0.00	2	0.49	–	0.44	0.04	0.00
500	EE	0	0.46	0.00	0.52	–	–	0	0.49	0.00	0.55	–	–
	IAMB	9	0.78	–	0.68	0.37	0.22	24	0.83	–	0.73	0.70	0.44
	IAMBFDR	6	0.68	–	0.59	0.02	0.00	7	0.70	–	0.60	0.02	0.00
1000	EE	0	0.68	0.00	0.70	–	–	0	0.70	0.00	0.73	–	–
	IAMB	14	0.84	–	0.74	0.35	0.23	27	0.88	–	0.78	0.68	0.46
	IAMBFDR	10	0.76	–	0.66	0.02	0.00	11	0.77	–	0.67	0.02	0.00

IAMB and IAMBFDR, we report $\overline{power}$, $\overline{FDR_1}$ and $\overline{FDR_2}$ which denote the average of $power_X$, $FDR_{X,1}$ and $FDR_{X,2}$ over all $X \in \mathbf{U}$. As discussed in Section 3, EE controls FDR, whereas IAMBFDR aims at controlling $\overline{FDR_2}$. We also report EE's $power$ for IAMB and IAMBFDR as well as $\overline{power}$ for EE, in order to assess the relative performance of the algorithms. Finally, we also report the runtimes of the algorithms in seconds (sec.). The runtimes correspond to C++ implementations of the algorithms run on a Pentium 2.0 GHz, 1 GB RAM and Windows XP.[2] We draw the following conclusions from the results in the table:

- As discussed above, EE is applicable only when $n > q$ which, as we will see in the next section, renders EE useless for learning GGMs of gene networks from most gene expression databases.
- In the cases where EE is applicable, EE controls FDR. This was expected as BY has been proven to control the FDR [3].
- IAMBFDR controls $\overline{FDR_2}$, though we currently lack a proof for this fact. IAMBFDR does not control $\overline{FDR_1}$, though it keeps it pretty low. The reason why IAMBFDR does not control $\overline{FDR_1}$ is in its iterative nature: If IAMBFDR fails to discover a node in $MB(X)$, then a node $Y \notin MB(X)$ may appear in the output MB of IAMB(X) or IAMBFDR(X). We think that this is a positive feature, as Y is informative about X because $X \not\perp Y | MB \setminus Y$ for Y to be included in MB. The average fraction of nodes in MB such that $Y \notin MB(X)$ but $X \not\perp Y | MB \setminus Y$ is $\overline{FDR_1} - \overline{FDR_2}$.
- IAMB controls neither $\overline{FDR_1}$ nor $\overline{FDR_2}$. As a matter of fact, the number of false discoveries made by IAMB(X) may get so large that the size of MB at line 3 exceeds $n - 3$, which implies that the hypothesis tests at that line cannot be run since the maximum likelihood estimates of the corresponding covariance matrices do not exist (recall Section 2). When this problem occurred, we aborted IAMB(X) and IAMB. With $\alpha = 0.05$, this problem occurred in the 100 databases with 50 samples, and in 26 databases with 100 samples. This problem also occurred when we applied IAMB to learn a GGM of a gene network from gene expression data (see next section), which compromises the use of IAMB for such a task.

---

[2] These implementations are available at www.ifm.liu.se/~jmp.

– IAMBFDR outperforms EE in terms of *power* whereas there is no clear winner in terms of $\overline{power}$. That IAMB outperforms the other two algorithms in terms of *power* and $\overline{power}$ is rather irrelevant, as it controls neither $\overline{FDR_1}$ nor $\overline{FDR_2}$. IAMBFDR is actually more powerful than what *power* and $\overline{power}$ indicate, as none of these measures takes into account the nodes $Y \in MB$ such that $Y \notin MB(X)$ but $X \not\perp Y | MB \setminus Y$ which, as discussed, above are informative about X.

In the light of the observations above, we conclude that IAMBFDR should be preferred to EE and IAMB: IAMBFDR offers FDR control while IAMB does not, moreover EE can only be run when $n > q$ in which case IAMBFDR is more powerful. Furthermore, the runtimes reported in Table 2 suggest that IAMBFDR scales to high-dimensional databases such as, for instance, gene expression databases. The next section confirms it. This is due to the fact that IAMBFDR exploits the composition property of Gaussian distributions to run forward, i.e. starting from the empty graph.

Finally, it is worth mentioning that we repeated all the experiments above with the unconditional means and the parental linear coefficients being uniformly drawn from [-1, 1], and the conditional standard deviations being equal to 1. The results obtained led us to the same conclusions as those above. As a sanity check, we also repeated all the experiments above with the sampled GGMs containing no edge. The results obtained confirmed that EE and IAMBFDR control the FDR even in such an extreme scenario whereas IAMB does not.

4.2   Rosetta Compendium

The Rosetta compendium [8] consists of 300 expression profiles of the yeast *Saccharomyces cerevisiae*, each containing expression levels for 6316 genes. Since for this database $n < q$, EE could not be run. Furthermore, the run of IAMB had to be aborted, since the problem discussed in the previous section occurred. Therefore, IAMBFDR was the only algorithm among those studied in this paper that could be run on the Rosetta compendium. Running IAMBFDR with Fisher's t-test and $\alpha = 0.01$ took 7.4 hours on a Pentium 2.4 GHz, 512 MB RAM and Windows 2000 (C++ implementation). The output contains 32641 edges, that is 0.16 % of all the possible edges.

In order to illustrate that the GGM learnt by IAMBFDR provides biological insight into the yeast gene network, we focus on the iron homeostasis pathway. Iron is an essential nutrient for virtually every organism, but it is also potentially toxic to cells. The iron homeostasis pathway regulates the uptake, storage, and utilization of iron so as to keep it at a non-toxic level. According to [12,20,21,23,24], yeast can use two different high-affinity mechanisms, reductive and non-reductive, to take up iron from the extracellular medium. The former mechanism is composed of the genes in the FRE family, responsible for iron reduction, and the iron transporters FTR1 and FET3, while the latter mechanisms consist of the iron transporters ARN1, ARN2, ARN3 and ARN4. The iron homeostasis pathway in yeast has been previously used in [16,17] to evaluate

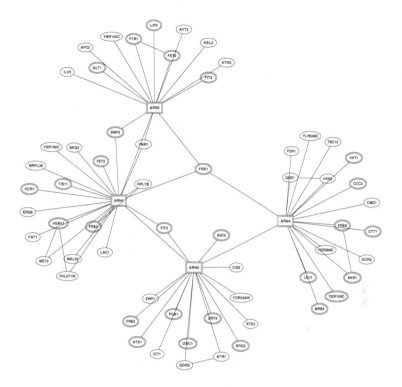

Fig. 1. Subgraph of GGM learnt by IAMBFDR that is induced by the genes that are adjacent to the four genes (square-shaped) involved in the non-reductive mechanism for iron uptake. Double-lined genes are related to iron homeostasis.

different algorithms for learning gene network models from gene expression data. These two papers conclude that their algorithms provide biologically plausible models of the iron homeostasis pathway after finding that many genes from that pathway are connected to ARN1 through a path of length one or two. We here take a similar approach to validate the GGM learnt.

Figure 1 depicts the subgraph of the GGM learnt that is induced by the genes that are adjacent to the four genes in the non-reductive mechanism for iron uptake, i.e. ARN1, ARN2, ARN3 and ARN4. These four genes are square-shaped in the figure. In addition to these, the figure contains many other genes related to iron homeostasis. These genes are double-lined in the figure. We now elaborate on these genes. As discussed above, FRE1, FRE2, FRE3, FRE6, FTR1 and FET3 are involved in the reductive mechanism for iron uptake. According to the Gene Ontology search engine AmiGO [1], FET5, MRS4 and SMF3 are iron transporters, FIT2 and FIT3 facilitate iron transport, PCA1 is involved in iron homeostasis, and ATX1 and CCC2 are involved in copper transport and are required by FET3, which is part of the reductive mechanism. According to [24], BIO2 is involved in biotin synthesis which is regulated by iron, GLT1 and ODC1 are involved in glutamate synthesis which is regulated by iron too, LIP5 is involved in lipoic acid synthesis and regulated by iron, and HEM15 is involved in heme synthesis and regulated by iron too. Also according to [24], TIS11 and the biotin transporter VHT1 are regulated by AFT1, the major iron-dependant transcription factor in yeast. Though AFT1 is not depicted in the subgraph in Figure 1, it is noteworthy that it is a neighbor of FET3 in the GGM learnt. The relation of the zinc transporter ZRT3 to iron homeostasis is documented in [23]. Finally, [5] provides statistical evidence that the following genes are related to iron homeostasis: LEU1, AKR1, HCR1, CTT1, ERG3 and YER156C. Besides, the paper confirms the relation of the first two genes through miniarray and quantitative PCR.

In summary, we have found evidence supporting the relation to iron homeostasis of 32 of the 64 genes in Figure 1. This means that, of the 60 genes that IAMBFDR linked to the four genes that we decided to study, 28 are related to iron homeostasis, which is a substantial fraction. Further evidence of the accuracy of the GGM learnt comes from the fact that these 60 genes are, according to the annotation tool g:Profiler [22], significantly enriched for several Gene Ontology terms that are related to iron homeostasis: GO:0055072 iron ion homeostasis (p-value $< 10^{-5}$), GO:0006825 copper ion transport (p-value $< 10^{-7}$), GO:0015891 siderophore transport (p-value $< 10^{-4}$), GO:0006826 iron ion transport (p-value $< 10^{-14}$), GO:0005506 iron ion binding (p-value $< 10^{-19}$), GO:0005507 copper ion binding (p-value $< 10^{-6}$), GO:0000293 ferric-chelate reductase activity (p-value $< 10^{-6}$), GO:0005375 copper ion transmembrane transporter activity (p-value $< 10^{-4}$), GO:0005381 iron ion transmembrane transporter activity (p-value $< 10^{-5}$), and GO:0043682 copper-transporting ATPase activity (p-value $= 10^{-4}$).

We think that the conclusions drawn in this section, together with those drawn in the previous section, prove that IAMBFDR is scalable and reliable for inferring GGMs of gene networks when $n < q$. Moreover, recall that neither EE nor IAMB could be run on the database used in this section.

5   Discussion

In this paper, we have proposed IAMBFDR, an algorithm for controlling the FDR when learning GGMs and $n < q$. We have shown that the algorithm works well in practice and scales to high-dimensional domains. In particulary, we have shown that IAMBFDR is able to provide biological insight in domains with thousands of genes but many fewer samples. Other works that propose algorithms for controlling the FDR when learning GGMs and $n < q$ are [25,26]. However, the correctness of the algorithm proposed in the first paper is neither proven nor fully supported by the experiments reported (e.g. see the results for sample size 50 in Figure 6 in [25]), whereas the algorithm in the second paper does not really aim at controlling the FDR but the closely related local FDR. IAMBFDR resembles the algorithms proposed in [6,14] in the sense that they all learn the GGM of a gene network by learning the MB of each node. Specifically, [6] takes a Bayesian approach that combines elements from regression and graphical models whereas [14] uses the lasso method. However, the main difference between our algorithm and theirs is that the latter do not aim at controlling the FDR. For the algorithm proposed in [14], this can clearly be seen in the experimental results reported in Table 1 in that work and in Figure 3 in [26].

Acknowledgements

This work is funded by the Swedish Research Council (ref. VR-621-2005-4202). We thank the anonymous reviewers for their insightful comments.

References

1. http://amigo.geneontology.org/cgi-bin/amigo/go.cgi
2. Anderson, T.W.: An Introduction to Multivariate Statistical Analysis. Wiley, Chichester (1984)
3. Benjamini, Y., Yekutieli, D.: The Control of the False Discovery Rate in Multiple Testing under Dependency. Annals of Statistics 29, 1165–1188 (2001)
4. Castelo, R., Roverato, A.: A Robust Procedure for Gaussian Graphical Model Search from Microarray Data with p Larger than n. Journal of Machine Learning Research 7, 2621–2650 (2006)
5. De Freitas, J.M., Kim, J.H., Poynton, H., Su, T., Wintz, H., Fox, T., Holman, P., Loguinov, A., Keles, S., van der Laan, M., Vulpe, C.: Exploratory and Confirmatory Gene Expression Profiling of *mac1Δ*. Journal of Biological Chemistry 279, 4450–4458 (2004)
6. Dobra, A., Hans, C., Jones, B., Nevins, J.R., Yao, G., West, M.: Sparse Graphical Models for Exploring Gene Expression Data. Journal of Multivariate Analysis 90, 196–212 (2004)
7. Geiger, D., Heckerman, D.: Learning Gaussian Networks. In: Proceedings of the Tenth Conference on Uncertainty in Artificial Intelligence, pp. 235–243 (1994)
8. Hughes, T.R., et al.: Functional Discovery via a Compendium of Expression Profiles. Cell 102, 109–126 (2000)
9. Jones, B., Carvalho, C., Dobra, A., Hans, C., Carter, C., West, M.: Experiments in Stochastic Computation for High Dimensional Graphical Models. Statistical Science 20, 388–400 (2005)
10. Kalisch, M., Bühlmann, P.: Estimating High-Dimensional Directed Acyclic Graphs with the PC-Algorithm. Journal of Machine Learning Research 8, 613–636 (2007)
11. Lauritzen, S.L.: Graphical Models. Oxford University Press, Oxford (1996)
12. Lesuisse, E., Blaiseau, P.L., Dancis, A., Camadro, J.M.: Siderophore Uptake and Use by the Yeast *Saccharomyces cerevisiae*. Microbiology 147, 289–298 (2001)
13. Meek, C.: Strong Completeness and Faithfulness in Bayesian Networks. In: Proceedings of the Eleventh Conference on Uncertainty in Artificial Intelligence, pp. 411–418 (1995)
14. Meinshausen, N., Bühlmann, P.: High-Dimensional Graphs and Variable Selection with the Lasso. Annals of Statistics 34, 1436–1462 (2006)
15. Pearl, J.: Probabilistic Reasoning in Intelligent Systems: Networks of Plausible Inference. Morgan Kaufmann, San Francisco (1988)
16. Pe'er, D., Regev, A., Elidan, G., Friedman, N.: Inferring Subnetworks from Perturbed Expression Profiles. Bioinformatics 224, S215–S224 (2001)

17. Peña, J.M., Nilsson, R., Björkegren, J., Tegnér, J.: Growing Bayesian Network Models of Gene Networks from Seed Genes. Bioinformatics 229, ii224–ii229 (2005)
18. Peña, J.M., Nilsson, R., Björkegren, J., Tegnér, J.: Reading Dependencies from the Minimal Undirected Independence Map of a Graphoid that Satisfies Weak Transitivity. In: Proceedings of the Third European Workshop on Probabilistic Graphical Models, pp. 247–254 (2006)
19. Peña, J.M., Nilsson, R., Björkegren, J., Tegnér, J.: Towards Scalable and Data Efficient Learning of Markov Boundaries. International Journal of Approximate Reasoning 45, 211–232 (2007)
20. Philpott, C.C., Protchenko, O., Kim, Y.W., Boretsky, Y., Shakoury-Elizeh, M.: The Response to Iron Deprivation in *Saccharomyces cerevisiae*: Expression of Siderophore-Based Systems of Iron Uptake. Biochemical Society Transactions 30, 698–702 (2002)
21. Protchenko, O., Ferea, T., Rashford, J., Tiedeman, J., Brown, P.O., Botstein, D., Philpott, C.C.: Three Cell Wall Mannoproteins Facilitate the Uptake of Iron in *Saccharomyces cerevisiae*. The Journal of Biological Chemistry 276, 49244–49250 (2001)
22. Reimand, J., Kull, M., Peterson, H., Hansen, J., Vilo, J.: g:Profiler – A Web-Based Toolset for Functional Profiling of Gene Lists from Large-Scale Experiments. Nucleic Acids Research 200, W193–W200 (2007)
23. Santos, R., Dancis, A., Eide, D., Camadro, J.M., Lesuisse, E.: Zinc Suppresses the Iron-Accumulation Phenotype of *Saccharomyces cerevisiae* Lacking the Yeast Frataxin Homologue (Yfh1). Biochemical Journal 375, 247–254 (2003)
24. Shakoury-Elizeh, M., Tiedeman, J., Rashford, J., Ferea, T., Demeter, J., Garcia, E., Rolfes, R., Brown, P.O., Botstein, D., Philpott, C.C.: Transcriptional Remodeling in Response to Iron Deprivation in *Saccharomyces cerevisiae*. Molecular Biology of the Cell 15, 1233–1243 (2004)
25. Schäfer, J., Strimmer, K.: An Empirical Bayes Approach to Inferring Large-Scale Gene Association Networks. Bioinformatics 21, 754–764 (2005)
26. Schäfer, J., Strimmer, K.: A Shrinkage Approach to Large-Scale Covariance Matrix Estimation and Implications for Functional Genomics. Statistical Applications in Genetics and Molecular Biology 4 (2005)
27. Studený, M.: Probabilistic Conditional Independence Structures. Springer, Heidelberg (2005)
28. Tsamardinos, I., Aliferis, C.F., Statnikov, A.: Algorithms for Large Scale Markov Blanket Discovery. In: Proceedings of the Sixteenth International Florida Artificial Intelligence Research Society Conference, pp. 376–380 (2003)
29. Werhli, A.V., Grzegorczyk, M., Husmeier, D.: Comparative Evaluation of Reverse Engineering Gene Regulatory Networks with Relevance Networks, Graphical Gaussian Models and Bayesian Networks. Bioinformatics 22, 2523–2531 (2006)
30. Whittaker, J.: Graphical Models in Applied Multivariate Statistics. John Wiley, Chichester (1990)
31. Wille, A., Bühlmann, P.: Low-Order Conditional Independence Graphs for Inferring Genetic Networks. Statistical Applications in Genetics and Molecular Biology 5 (2006)
32. Wille, A., Zimmermann, P., Vranova, E., Fürholz, A., Laule, O., Bleuler, S., Hennig, L., Prelic, A., von Rohr, P., Thiele, L., Zitzler, E., Gruissem, W., Bühlmann, P.: Sparse Graphical Gaussian Modeling of the Isoprenoid Gene Network in *Arabidopsis thaliana*. Genome Biology 5, 1–13 (2004)

Enhancing Parameter Estimation of Biochemical Networks by Exponentially Scaled Search Steps

Hendrik Rohn, Bashar Ibrahim, Thorsten Lenser,
Thomas Hinze, and Peter Dittrich

Jena Centre for Bioinformatics
and
Friedrich Schiller University Jena
Bio Systems Analysis Group
Ernst-Abbe-Platz 1–4, D-07743 Jena, Germany
{morla,ibrahim,thlenser,hinze,dittrich}@minet.uni-jena.de

Abstract. A fundamental problem of modelling in Systems Biology is
to precisely characterise quantitative parameters, which are hard to measure experimentally. For this reason, it is common practise to estimate
these parameter values, using evolutionary and other techniques, by fitting the model behaviour to given data. In this contribution, we extensively investigate the influence of exponentially scaled search steps on the
performance of two evolutionary and one deterministic technique; namely
CMA-Evolution Strategy, Differential Evolution, and the Hooke-Jeeves
algorithm, respectively. We find that in most test cases, exponential scaling of search steps significantly improves the search performance for all
three methods.

1 Introduction

At the beginning of the 21st century, the area of Systems Biology has a major
and still widening impact on the future of biological and medical research [15,16].
Computational models in Systems Biology often have numerous parameters,
such as kinetic parameters (e.g. reaction rates), saturation constants and dif-
fusion constants. Directly measuring unknown biochemical parameters *in vivo*
is difficult, and collectively fitting them to other experimental data often yields
large parameter uncertainties. As a result, methods for the estimation of these
parameters are central and of great importance to this field of research.

For this task, evolutionary approaches (e.g. [12]) as well as deterministic algorithms (e.g. Hooke-Jeeves [8]) have been applied to infer parameters in biochemical models. Although these approaches are among the most successful in
their class, it is very difficult to estimate the parameters when there are many
interactions in the system under consideration, having many local optima. They
consume enormous computational time because they require iterative numerical
integrations for non-linear differential equations. When the target model is stiff,
the computational time for reaching a solution increases further. Moreover, the

E. Marchiori and J.H. Moore (Eds.): EvoBIO 2008, LNCS 4973, pp. 177–187, 2008.

difficulty of the problem increases not only with the number of parameters, but also with the width of the search interval for each parameter.

Recent biological problems, from yeast to human systems, contain many free parameters to fit (see for example [17,24]), and the parameter space varies widely [13,21,28]. Optimisation techniques often do not work efficiently and accurately without extrinsic mathematical hints from the user, like parameter space scaling.

In the work presented here, we demonstrate that parameter estimation can be efficiently and more accurately performed using scaled search steps, in analogy to the "log-normal rule" of adapting the mutational variance in evolution strategies [1]. This idea is supported by testing different networks taken from the BioModels database [25], using three different optimisation techniques. Two of them are evolutionary approaches; Covariance Matrix Adaptation Evolutionary Strategy (CMA-ES) [5], and Differential Evolution (DE) [27]. The third one represents a deterministic optimisation, the method of Hooke and Jeeves (HJ) [8].

2   Selected Optimisation Algorithms

Evolution Strategies (ES) [6,7] are stochastic search algorithms which try to minimize a given objective function. The search steps are realized by stochastic variation, often called mutations due to the comparable biological events. After starting with a random set of search points (often called individuals) these points are evaluated under usage of a given objective function. A small set of good performing search points is chosen and varied by stochastic combination and variation, representing the next step of the algorithm. Now the set of search points is evaluated and some are selected to be the origin of the new set. This approach guarantees a movement in the search space to an (local) optimum.

Usually, no detailed knowledge of adequate settings of the mutation-parameters is available, which is crucial to the algorithm's performance on a given problem. Covariance Matrix Adaptation[1] uses information collected during the search process to select a suitable set of mutation-rates for the problem at hand. It is a second-level derandomized self-adaptation scheme and directly implements the Mutative Strategy Parameter Control [7].

Differential Evolution[2] (DE) [26,27] represents a second evolutionary approach and thereby a stochastic search algorithm. It is often mentioned in the context of Genetic Algorithms and Evolution Strategies, but it has some distinct properties. The procedure also uses populations consisting of vectors, which are potential solutions to the optimisation problem. Like Evolutionary Algorithms the first population is drawn randomly. DE generates new vectors by adding the weighted difference between two vectors to a third vector, which is called mutation. It also mixes this resulting vector and a previously determined other,

---

[1] Implementation taken from *bionik.tu-berlin.de/user/niko/cmaes_inmatlab.html*, parameter settings: insigma=0.3(upper_bound-lower_bound), PopSize=3+floor (3log(D)), Restarts=0.

[2] Implementation taken from *icsi.berkeley.edu/~storn/code.html#matl* parameter settings: NP = 10D, itermax=5.

so-called target vector, together, which results in a trial-vector in a new trial-population. Each vector has to serve once as a target vector and thereby fills the new trial-population. The selection scheme also differs from Evolutionary Algorithms (EA): each vector of the trial-population is compared with its counterpart in the current generation, and the better one survives into the next generation. This means that each individual of the new generation is the same or better than the corresponding individual from the old population. Therefore, the difference to typical EA selection schemes is that a trial-vector does only have to compete against its predecessor, not against all other vectors.

On some problems, DE outperforms several deterministic and stochastic approaches in convergence speed and finding the global optimum, including some Evolutionary Algorithms [27]. The speed can be topped only by some deterministic approaches, but they are only applicable to some special problems.

As a representative for a deterministic search procedure, the algorithm[3] proposed by Hooke and Jeeves (HJ) [8,14] is often referred to as a "direct search" method. The algorithm starts with a single point, called base point, sometimes randomly drawn. Then a second search point, a trial-point, is chosen and compared to the base point. If the new point represents a better solution, it is taken as the second base point, otherwise it is neglected and the old point remains the base. The process of choosing and comparing continues until a stop criterion is reached. The strategy for selecting new trial-points is based on two moves: exploratory and pattern moves.

Exploratory moves are usually simple, like changing only one dimension of the search point, which are repeated to investigate the surrounding landscape. They are calculated by changing each coordinate separately by a certain step-size and investigating the failure or success. For a success the change will be kept, otherwise restored. The last move of one step will then be the pattern move. It thereby combines the gathered information into a directed move in a probably good direction of the search space and generates a new base vector. If there is no better trial-point discovered after a certain amount of failed pattern moves the step size will be reduced to be able to converge more slowly to the optimum. Termination finally happens when the step size is small enough, assuming that the movement to a minimum has been successful. The approach allows to take into account more previous base points, to determine a global success despite local failure in the search direction, which can improve the search process. In practical cases direct search is successful in locating minima in "sharp valleys", because each (exploratory) step is only a minor step in the search space.

3   Networks

In the presented experiments, 12 different networks were considered, of which all but one are available in the BioModels Database [25]: Fisher et al. [2] model NFAT and NFκ-B activation in T lymphocytes. Fung et al. [4] describe a

---

[3] Implementation taken from *mathworks.com/matlabcentral/fileexchange/load File.do?objectId* − 15070 parameter settings: tol=10^{-4}, mxit=problem dimension.

synthetic gene-metabolic oscillator. Hornberg et al. [9] describe ERK phosphorylation and kinase/phosphatase control. Huang and Ferrell [10] show the ultra-sensitivity in the mitogen-acitivated protein kinase cascade. Kofahl and Klipp [17] model the dynamics of the yeast pheromone pathway. Kongas and van Beek [18] delineate the energy metabolic signaling in muscles through creatine kinase, and Martins and van Boekel [20] model the Amadori degradation pathway. Marwan [21] discusses the sporulation control network in *Physarum polycephalum*. Nielsen et al. [24] study sustained oscillations in glycolysis and its complex periodic behaviour. In a widely recognized study, Tyson [28] analysed the cell division cycle, and Yildirim and Mackey [30] examined a feedback regulation in the lactose operon. The last network is an artificial one that can calculate the third root of a positive number, i.e. the concentration of the output-species is the third root of the concentration of the one input-species [19]. Thus, using all three optimisation algorithms for all 12 networks, a set of 36 parameter fitting test-cases has been investigated. Table 2 contains the network names together with the network properties.

The networks, represented in SBML (Systems Biology Markup Language [11]), were converted into ordinary differential equation (ODE) systems and stored as MATLAB m-files [22] using CellDesigner [3]. The integration of the ODEs was done by the built-in MATLAB-function ode15s. Additionally, COPASI [23] has been used to integrate and visualize the behaviour of all species in each network. Subsequently, a small number of species of each network has been chosen, which showed an interesting behaviour, like oscillations and peaks. These time-series were then defined as the "optimal" behaviour of a network, such that the algorithms start with random parameter sets and have to optimise the parameters of the network, to produce the same output as the original parameterized network. As an objective function, the quadratic distance between the output of the original and the candidate-network was used. Therefore, a parameter set is good if the behaviour of the network is similiar to the original network, which usually is the case for parameters similiar to the original ones (as well as for other parameter sets, since parameter fitting is commomly an underdetermined problem). Accordingly, a parameter set is bad if the species show different behaviour.

The networks were allowed to have parameters in the range between 10^{-5} and 10^5, containing most of the desired parameters of each network. The implementation by Hooke and Jeeves and Differential Evolution didn't support constraint parameters, so the implementations had to be extended: each time a new parameter set is generated, each parameter is checked for a bound violation and set back to the bound if it exceeds it.

4   Scaling of Parameters

The task of the work presented here is to investigate a possible improvement of the optimisation procedures by scaled search steps. The optimisation algorithms in the unscaled case allow parameters p_i in the range between 10^{-5} to 10^5, the scaled case considers parameters p_i between 0 to 1. In the unscaled case, the

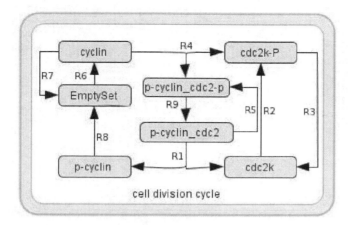

Fig. 1. The network proposed by Tyson [28], visualised with CellDesigner [3]

unchanged parameters p_i were inserted into the ODEs and evaluated by integration and measuring the distance between the derived and the original behaviour. In the scaled case, the changed parameters $\bar{p}_i$ were inserted and evaluated. The relationship between p_i and $\bar{p}_i$ is

$$\bar{p}_i = a \cdot 10^{b+c \cdot p_i},$$

where $a = 1$, $b = -5$ and $c = 10$ resulting in $10^{-5} \leq \bar{p}_i \leq 10^5$. As a result of this scaling, search steps are larger for high parameter values and smaller for low values, independent of the optimisation procedure. This allows a searching procedure to investigate small parameter values with high resolution and large parameter values with low resolution without having to internally adapt the search step-size.

A good example is the cell division cycle model proposed by Tyson [28], which is shown in Figure 1. It consists of 6 species and 9 reactions, resulting in 6 ordinary differential equations with the parameters

- R1_k6 = 1.0
- R2_k8notP = 1000000.0
- R3_k9 = 1000.0
- R4_k3 = 200.0
- R5_k5notP = 0.0
- R6_k1aa = 0.015
- R7_k2 = 0.0
- R8_k7 = 0.6
- R9_k4 = 180.0
- R9_k4prime = 0.018

It can clearly be seen that the second parameters' magnitude is larger than all other parameters. To show the advantage of the scaled approach, let us assume that the step-size s during the search for each parameter is small in the search space, e.g.

$$s = 10^{-4} \cdot (\text{upperbound} - \text{lowerbound}).$$

This results in a step-size of $s_{\text{unscaled}} = 10$ and $s_{\text{scaled}} = 10^{-4}$, respectively. Now assume that during the search

$$\text{R2\_k8notP} = 90000 \approx 10^{4.5424}$$

and

$$\text{R6\_k1aa} = 0.5 \approx 10^{-0.30103}.$$

The next optimisation step would generate new unscaled parameters

$$\text{R2\_k8notP} = 90000 + 10 = 90010$$

and

$$\text{R6\_k1aa} = 0.5 - 10 = -9.5$$

(which would also violate the bounds). The change is the same for both parameter regions, but has a higher impact on the sixth parameter than on the second. With the scaled approach, new parameters

$$\text{R2\_k8notP} = 10^{4.954254+10^{-4}} = 90023.10745$$

and

$$\text{R6\_k1aa} = 10^{-0.301-10^{-4}} = 0.49992$$

are obtained, which is a more intelligent way to search the landscape for biological networks. The biological background is that the parameters of a network model usually have different meanings. They can be grow constants or saturation constants, which have different units: grow constants are rates (per molecule or similar) and thereby have low orders of magnitude. Saturation constants have the unit molecule number (or concentration), which is in higher orders of magnitude. Since the optimisation procedures are desired to be as general as possible, the scale of each parameter is not known a priori. The scaled mutations proposed here amend this situation.

5   Results

For each network, each optimisation algorithm was run 50 times, and the quality of the solution together with the speed of reaching it was recorded. We define the quality as the value of the objective function at the end of the run, while we take the average value of the objective function over the whole run (except for the initial phase which depends on the starting points) as a measure of speed. Figure 2 contains six example runs. Table 1 shows a compact overview of the results, and Table 2 gives more detailed information including the measured values for quality and speed.

In Table 1, the first column gives the name of the main author of the network. Second and third column indicate the better (plus) or worse performance (minus) of the scaled approach compared to the unscaled for the final reached objective function value ("better") and the mean value over all function evaluations

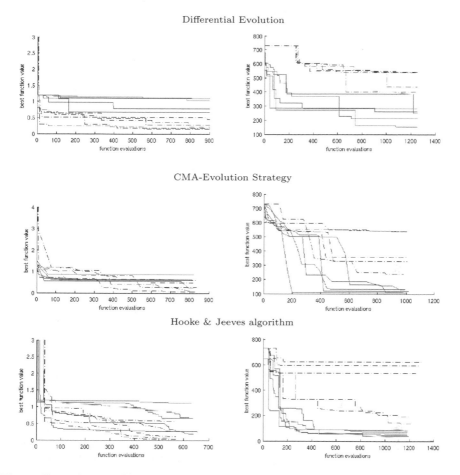

Fig. 2. Typical runs of six exemplary fitting cases of the networks by Hornberg and Nielsen, optimised by all three algorithms; the dashed lines indicate the unscaled approach, solid lines show the scaled one

("faster"), achieved by Differential Evolution. Brackets mark non-significant results according to the Mann-Whitney-Wilcoxon-Test [29], with P-values larger than 10^{-3}. The next four columns show the same for the CMA-Evolution Strategy and the Hooke and Jeeves algorithm.

Table 1 clearly points out that the scaled approach outperforms the unscaled one in most cases, as expected. Some test-cases show only a small difference (Fisher, Fung, third root) and a few give worse results (Hornberg, Huang). All other test-cases show an improvement in speed and/or solution quality for the optimisation procedure. The reason for the pronounced difference in the effect of scaled search steps still remains to be examined. One likely point is that the behaviour of these networks might be quite sensitive to changes in the large

Table 1. Advantage (plus) or disadvantage (minus) of the scaled approach, brackets indicate a non-significant result at the 10^{-3} level. Based on 50 indepedent runs for each test-case.

author	—DE— better	faster	—ES— better	faster	—HJ— better	faster
Fisher [2]	-	-	(-)	(-)	(+)	+
Fung [4]	-	+	(+)	(+)	+	+
Hornberg [9]	-	-	-	-	-	-
Huang [10]	-	-	-	-	-	-
Kofahl [17]	+	+	+	+	+	+
Kongas [18]	(+)	+	+	+	+	+
Martins [20]	+	+	+	+	+	(+)
Marwan [21]	+	+	+	+	+	+
Nielsen [24]	+	+	+	+	+	+
third root [19]	-	(-)	-	(-)	(-)	(-)
Tyson [28]	+	+	+	+	-	(+)
Yildirm [30]	+	+	+	+	+	+

parameters, in which case the scaling may prevent adequate fine-tuning of these parameters. However, it also has to be noted that in cases where performance is worse, this effect is not too pronounced, giving more weight to the partially drastic performance increase in most other cases.

6   Conclusions

Evidently, the performance of parameter fitting algorithms depends on the way in which the search steps are carried out. In this paper, we propose an exponential scaling of these steps, which leads to larger changes for larger parameters and smaller ones for smaller parameters. To evaluate the effect of the proposed scaling technique on fitting performance, we downloaded 12 models from the BioModels database and seeded them with random parameters. The fitting algorithms were then used to re-create the original network's behaviour. To show that the advantage of exponentially scaled search steps is independent of the search algorithm used, we employed three different optimisation techniques in this study.

The experimental results (Table 2) clearly confirm an advantage of using scaled search steps, in particular when the search interval for the parameters is large. When the interval to be searched is not too large anyway, our results show that exponential scaling at least does not deteriorate performance in most cases. Thus, when fitting parameters of today's system biological models, scaling the search steps exponentially can be recommended as a default approach.

Table 2. Overview of all results, based on 50 repetitions per test-case; "optim." = optimisation algorithm, "nbr" = number of network's parameters, "var" = statistical variance of the original parameters, ratio = $\frac{\text{upperbound}}{10^5 \cdot \text{lowerbound}}$, "better" and "faster" = advantage of the scaled approach for the final objective function value and optimisation speed respectively (brackets indicate non-significant results), "final f-value" and "mean f-value" = numerical results for the unscaled and scaled approach, "significance" = P-value of the rank-sum test [29] (not significant, if $P > 10^{-3}$), "*" indicates where outliers cause pronounced differences in mean values despite non-significant differences in distributions

author	optim.	nbr	var	ratio	better	faster	final f-value			mean f-value		
							unscaled	scaled	significance	unscaled	scaled	significance
Fisher	DE	22	3.64	2.59	-	-	0.012712	0.01459	2.964e-4	0.013786	0.014668	2.8624e-4
Fisher	ES	22	3.64	2.59	(-)	(-)	0.014392	0.014567	4.7558e-3	0.015063	0.01516	9.0598e-3
Fisher	HJ	22	3.64	2.59	(+)	+	0.014745	0.014718	3.8323e-1	0.014773	0.014749	1.7417e-5
Fung	DE	22	691.54	800	-	+	1680.7444	1881.6649	1.5444e-12	2.1349e7	1922.8123	6.508e-18
Fung	ES	22	691.54	800	(+)	(+)	2e9*	1802.6033	3.4288e-3	2e9*	1983.7889	7.5872e-2
Fung	HJ	22	691.54	800	+	+	5e10	1e10	2.8657e-4	5.0180e10	1.0019e10	1.3275e-13
Hornberg	DE	18	0.2	0.001	-	-	0.24796	0.69365	1.5146e-16	0.41718	0.97441	8.9852e-18
Hornberg	ES	18	0.2	0.001	-	-	0.24796	0.69365	5.5646e-12	0.43934	0.95489	3.5813e-16
Hornberg	HJ	18	0.2	0.001	-	-	0.29467	0.48748	1.9984e-4	0.51723	0.74825	6.1039e-5
Huang	DE	30	1.66e5	6.66e-5	-	-	14.403	16.0303	1.01e-16	14.887	16.0746	1.01e-16
Huang	ES	30	1.66e5	6.66e-5	-	-	14.607	15.6329	2.4811e-8	15.234	16.25	9.305e-11
Huang	HJ	30	1.66e5	6.66e-5	-	-	15.655	16.0946	1.1825e-7	15.7575	16.1244	1.3665e-4
Kofahl	DE	47	8.66e4	16.67	-	+	2.4844e7	1.8342e7	1.3657e-17	3.9667e7	1.9215e7	7.0661e-18
Kofahl	ES	47	8.66e4	16.67	+	+	3.2578e7	1.8611e7	3.7251e-11	3.6849e7	1.9545e7	1.1018e-13
Kofahl	HJ	47	8.66e4	16.67	+	+	7.3459e7	2.0194e7	5.5337e-15	7.4303e7	2.0436e7	1.6045e-16
Kongas	DE	25	1.03e8	0.04	(+)	+	1.7501e11	1.1418e11	9.9508e-3	3.0034e11	2.1588e11	8.1319e-5
Kongas	ES	25	1.03e8	0.04	+	+	3.5567e11	2.3476e11	9.8645e-5	3.679e11	2.7196e11	6.9503e-4
Kongas	HJ	25	1.03e8	0.04	+	+	4.5529e11	3.8867e11	1.7231e-8	4.5568e11	4.0287e11	3.7706e-12
Martins	DE	16	0.23	0.02	+	+	129.2063	40.6938	7.0661e-18	129.6498	58.504	7.0661e-18
Martins	ES	16	0.23	0.02	+	+	129.2157	50.5015	7.0661e-18	136.1664	76.4991	2.5415e-16
Martins	HJ	16	0.23	0.02	(+)	+	137.3989	84.9211	1.4934e-7	138.3261	99.6395	2.8386e-4
Marwan	DE	10	302.53	0.01	+	+	199016	34679	7.5041e-18	1.0242e9	53861	7.0661e-18
Marwan	ES	10	302.53	0.01	+	+	182076	33307	7.0661e-18	5.7994e10	65460	3.7392e-17
Marwan	HJ	10	302.53	0.01	+	+	293190	65665	3.5664e-15	1.4741e11	188174	7.5041e-13
Nielsen	DE	25	104.48	3.5	+	+	535.7951	228.0287	7.0661e-18	579.1766	286.8025	7.0661e-18
Nielsen	ES	25	104.48	3.5	+	+	368.3119	159.0676	1.3213e-11	472.8056	263.5361	5.0161e-14
Nielsen	HJ	25	104.48	3.5	+	+	505.5813	113.414	1.7701e-5	529.3175	164.0565	5.6428e-16
third root	DE	6	0.81	1400	-	-	11.0738	42.2334	9.8191e-5	121.685	165.1885	7.252e-2
third root	ES	6	0.81	1400	(-)	(-)	0.12539	13.2355	4.388e-5	80.3153	125.0264	6.4171e-2
third root	HJ	6	0.81	1400	(-)	(-)	212.2248	266.478	7.4741e-2	758.2694	31331.085*	2.5678e-1
Tyson	DE	10	e11	66.66	+	+	122.9164	120.1576	9.3754e-16	135.5054	121.3539	7.0661e-18
Tyson	ES	10	e11	66.66	+	+	120.812	121.1046	7.0311e-18	133.9525	129.0575	1.7755e-5
Tyson	HJ	10	e11	66.66	(+)	+	126.4592	131.1673	2.0426e-15	145.2656	145.2234	8.4185e-2
Yildirim	DE	23	5.53e7	3.47e5	+	+	6.1058	4.6403	1.1283e-13	8.3897	5.2105	7.0661e-18
Yildirim	ES	23	5.53e7	3.47e5	+	+	9.6389	3.6808	4.8399e-13	10.4665	4.8209	1.0357e-12
Yildirim	HJ	23	5.53e7	3.47e5	+	+	13.1821	6.1579	7.3666e-16	13.2274	6.7414	3.3014e-20

Acknowledgements

Funding from the EU (ESIGNET, project no. 12789), Federal Ministry of Education and Research (BMBF, grant 0312704A) and German Academic Exchange Service (DAAD, grant A/04/31166) is gratefully acknowledged.

References

1. Beyer, H.-G., Schwefel, H.-P.: Evolution strategies. Natural Computing 1, 3–52 (2002)
2. Fisher, W.G., Yang, P.C., Medikonduri, R.K., Jafri, M.S.: NFAT and NFκB activation in T lymphocytes: a model of differential activation of gene expression. Ann Biomed Eng 34(11), 1712–1728 (2006)
3. Funahashi, A., Tanimura, N., M.M., Kitano, H.: CellDesigner: A process diagram editor for gene-regulatory and biochemical networks. BIOSILICO 1, 159–162 (2003)
4. Fung, E., Wong, W.W., Suen, J.K., Bulter, T., Lee, S., Liao, J.C.: A synthetic gene-metabolic oscillator. Nature 435(7038), 118–122 (2005)
5. Hansen, N., Kern, S.: Evaluating the cma evolution strategy on multimodal test functions. In: Eighth International Conference on Parallel Problem Solving from Nature PPSN VIII, pp. 282–291. Springer, Heidelberg (2004)
6. Hansen, N., Muller, S.D., Koumoutsakos, P.: Reducing the time complexity of the derandomized evolution strategy with covariance matrix adaptation (CMA-ES). Evol Comput 11(1), 1–18 (2003)
7. Hansen, N., Ostermeier, A.: Completely derandomized self-adaptation in evolution strategies. Evol Comput 9(2), 159–195 (2001)
8. Hooke, R., Jeeves, T.A.: " direct search" solution of numerical and statistical problems. J. ACM 8(2), 212–229 (1961)
9. Hornberg, J.J., Bruggeman, F.J., Binder, B., Geest, C.R., de Vaate, A.J.M.B., Lankelma, J., Heinrich, R., Westerhoff, H.V.: Principles behind the multifarious control of signal transduction. ERK phosphorylation and kinase/phosphatase control 272(1), 244–258 (2005)
10. Huang, C.Y., Ferrell, J.E.J.: Ultrasensitivity in the mitogen-activated protein kinase cascade. Proc Natl Acad Sci USA 93(19), 10078–10083 (1996)
11. Hucka, M., Finney, A., Bornstein, B.J., Keating, S.M., Shapiro, B.E., Matthews, J., Kovitz, B.L., Schilstra, M.J., Funahashi, A., Doyle, J.C., Kitano, H.: Evolving a lingua franca and associated software infrastructure for computational systems biology: The systems biology markup language (SBML) project. Systems Biology 1(1), 41–53 (2004)
12. Ibrahim, B., Diekmann, S., Schmitt, E., Dittrich, P.: *In-silico* model of the mitotic spindle assembly checkpoint. PLoS one (Under revision 2008)
13. Ibrahim, B., Schmitt, E., Dittrich, P., Diekmann, S.: MCC assembly is not combined with full Cdc20 sequestering (Submitted 2007)
14. Kaupe Jr., A.F.: Algorithm 178: Direct search. Commun. ACM 6(6), 313–314 (1963)
15. Kitano, H.: Computational systems biology. Nature 420(14), 206–210 (2002)
16. Kitano, H.: Systems biology: a brief overview. Science 295(5560), 1662–1664 (2002)
17. Kofahl, B., Klipp, E.: Modelling the dynamics of the yeast pheromone pathway. Yeast 21(10), 831–850 (2004)

18. Kongas, O., van Beek, J.H.G.M.: Creatine kinase in energy metabolic signaling in muscle. In: Proc. 2nd Int. Conf. Systems Biology (ICSB 2001), pp. 198–207 (2001)
19. Lenser, T., Hinze, T., Ibrahim, B., Dittrich, P.: Towards evolutionary network reconstruction tools for systems biology. In: Marchiori, E., Moore, J.H., Rajapakse, J.C. (eds.) EvoBIO 2007. LNCS, vol. 4447, Springer, Heidelberg (2007)
20. Martins, S.I.F.S., Boekel, M.A.J.S.V.: Kinetic modelling of Amadori N-(1-deoxy-D-fructos-1-yl)-glycine degradation pathways. Part II–kinetic analysis. Carbohydr Res 338(16), 1665–1678 (2003)
21. Marwan, W.: Theory of time-resolved somatic complementation and its use to explore the sporulation control network in Physarum polycephalum. Genetics 164(1), 105–115 (2003)
22. Mathworks: (Retrieved June 20, 2007) (2007), http://www.mathworks.com/
23. Mendes group at VBI and Kummer group at EML research. COPASI: (Retrieved June 20, 2007) (2007), http://www.copasi.org/
24. Nielsen, K., Sorensen, P.G., Hynne, F., Busse, H.G.: Sustained oscillations in glycolysis: An experimental and theoretical study of chaotic and complex periodic behavior and of quenching of simple oscillations. Biophys Chem 72(1–2), 49–62 (1998)
25. Novre, N.L., Bornstein, B., Broicher, A., Courtot, M., Donizelli, M., Dharuri, H., Li, L., Sauro, H., Schilstra, M., Shapiro, B.,, J.S.L., Hucka, M.: BioModels database: A free, centralized database of curated, published, quantitative kinetic models of biochemical and cellular systems. Nucleic Acids Research 34 (2006)
26. Storn, R.: On the usage of differential evolution for function optimization. In: Biennial Conference of the North American Fuzzy Information Processing Society, NAFIPS, pp. 519–523 (1996)
27. Storn, R., Price, K.: Differential evolution - a simple and efficient heuristic for global optimization over continuous spaces. J. of Global Optimization 11(4), 341–359 (1997)
28. Tyson, J.J.: Modeling the cell division cycle: cdc2 and cyclin interactions. Proc Natl Acad Sci USA 88(16), 7328–7332 (1991)
29. Wilcoxon, F.: Individual comparisons by ranking methods. Biometrics Bulletin 1, 80–83 (1945)
30. Yildirim, N., Mackey, M.C.: Feedback regulation in the lactose operon: A mathematical modeling study and comparison with experimental data. Biophys J 84(5), 2841–2851 (2003)

A Wrapper-Based Feature Selection Method for ADMET Prediction Using Evolutionary Computing

Axel J. Soto[1,2], Rocío L. Cecchini[1], Gustavo E. Vazquez[1], and Ignacio Ponzoni[1,2]

[1] Laboratorio de Investigación y Desarrollo en Computación Científica (LIDeCC),
Departamento de Ciencias e Ingeniería de la Computación (DCIC)
Universidad Nacional del Sur – Av. Alem 1253 – 8000 – Bahía Blanca
Argentina
[2] Planta Piloto de Ingeniería Química (PLAPIQUI)
Universidad Nacional del Sur – CONICET
Complejo CRIBABB – Camino La Carrindanga km.7 – CC 717 – Bahía Blanca
Argentina
{saj,rlc,gev,ip}@cs.uns.edu.ar

Abstract. Wrapper methods look for the selection of a subset of features or variables in a data set, in such a way that these features are the most relevant for predicting a target value. In chemoinformatics context, the determination of the most significant set of descriptors is of great importance due to their contribution for improving ADMET prediction models. In this paper, a comprehensive analysis of descriptor selection aimed to physicochemical property prediction is presented. In addition, we propose an evolutionary approach where different fitness functions are compared. The comparison consists in establishing which method selects the subset of descriptors that best predicts a given property, as well as maintaining the cardinality of the subset to a minimum. The performance of the proposal was assessed for predicting hydrophobicity, using an ensemble of neural networks for the prediction task. The results showed that the evolutionary approach using a non linear fitness function constitutes a novel and a promising technique for this bioinformatic application.

Keywords: Feature Selection, Genetic Algorithms, QSAR, hydrophobicity.

1 Motivation

In the pharmaceutical industry, when a new medicine has to be developed, a 'serial' process starts where drug potency (activity) and selectivity are examined first [1]. Many of the candidate compounds fail at later stages due to ADMET (absorption, distribution, metabolism, excretion and toxicity) behavior in the body. ADMET properties are related to the way that a drug interacts with a large number of macromolecules and they correspond to the principal cause of failure in drug development [1]. In this way, a compound can be promising at first based on its molecular structure, but other factors such as aggregation, limited solubility or limited uptake in the human organism turn it useless as a drug.

Nowadays, the failure rate of a potential drug before reaching the market is still high. The main problem is that most of the rules that govern ADMET behavior in the

E. Marchiori and J.H. Moore (Eds.): EvoBIO 2008, LNCS 4973, pp. 188–199, 2008.

human body are unknown. For these reasons, interest in Quantitative Structure-Activity Relationships (QSAR) and Quantitative Structure-Property Relationships (QSPR) given by the scientific and industrial community has grown considerably in the last decades. Both of these approaches comprise the methods by which chemical structure parameters (known as descriptors) are quantitatively correlated with a well defined process, such as biological activity or any other experiment. QSAR has evolved over a period of 30 years from simple regression models to different computational intelligence models that are now applied to a wide range of problems [2], [3]. Nevertheless, the accuracy of the ADMET property estimations remains as a challenging problem [4].

In this context, hydrophobicity is one of the most extensively modeled physicochemical properties since the difficulty of experimentally determine its value, and also because it is directly related to ADMET properties [2], [5]. This property is traditionally expressed in terms of the logarithm of the octanol-water partition coefficient (logP).

QSAR methods developed by computer means are commonly named as *in silico* methods. These *in silico* methods, clearly cheaper than *in vitro* experiments, allow to examine thousands of molecules in shorter time and without the necessity of intensive laboratory work. Although *in silico* methods are not pretended to replace high-quality experiments at least in the short term, some computer methods have demonstrated to obtain as good accuracy as well-established experimental methods [6]. Moreover, one of the most important features of this approach is that a candidate drug (or a whole library) can be tested before being synthesized. Due to the gains in saved labour time, *in silico* predictions considerably help to reduce the large percentage of leads that fail in later stages of their development, and to avoid the amount of time and money invested in compounds that will not be successful.

In this context, machine learning methods are most preferred given the great amount of existing data and the little understanding of the pharmacokinetic rules of xenobiotics in the human body. Jónsdottir *et al.* [3] detail an extensive review of the many machine learning methods applied to bio- and chemoinformatics.

The major dilemma when logP is intended to be modeled by QSAR is that, thousands of descriptors could be measured for a single compound and also there is no general agreement on which descriptors are relevant or influence the hydrophobic behavior of a compound. This is an important fact, because overfitting and chance correlation could occur as a result of using more descriptors than necessary [7], [8]. On the other hand, poor models come as a result, when less descriptors than necessary are used. From an Artificial Intelligence (AI) perspective, this topic constitutes a particular case of the feature selection (FS) problem.

In this way, this work presents a sound approach for inferring the subset of the most influential descriptors for physicochemical properties. The righteousness of the selection is assessed by the construction of a prediction model. Our technique is based in the application of a genetic algorithm (GA) where: different fitness functions, a different number of descriptors selected by GA and a different number of descriptors considered by the prediction method are compared. This work is organized as follows: next section discusses related issues of feature selection in AI and in chemoinformatics in particular. Section 3 expands the aforementioned idea by introducing the genetic algorithm proposed for descriptor selection. In Section 4, applied data and methods are presented, followed by the obtained results. Finally, in Section 5, main conclusions and future work are discussed.

2 Introduction to Feature Selection

Feature selection is the common name used to comprise all the methods that select from or reduce the set of variables or features used to describe any situation or activity in a dataset. Some authors differentiate variables from features, assuming that variables are the raw entry data, whereas features correspond to processed variables. However, variables, features or descriptors will be used here without distinction.

Nowadays, FS is a current research area, given that applications with datasets of many (even hundreds or thousands) variables have become frequent. Most usual cases where this technique is applied are gene selection from microarray data [9], [10], [11] and text categorization [12], [13], [14]. Confronting dimensionality carries some recognized advantages like: reducing the measurement and storage requirements, facilitating visualization and understanding of data, diminishing training and predicting times and also improving prediction performance.

Special care has to be taken with the distinction between relevant or useful and redundant. As it can be elucidated, selecting most relevant variables may be suboptimal for a predictor, especially when relevant variables are redundant. On the other hand, a subset of useful variables for a predictor may exclude redundant, but relevant, variables [15], [16], [17]. Therefore, in FS it is important to know whether developing a predictor is a final objective or not.

FS methods may be applied in two main ways, in terms of whether variables are individually or globally evaluated. That is, the first of them, works ranking each variable in an isolated way, i.e. these methods rank variables according to their individual predictive power. However, a variable that is useless by itself could be useful in consideration with others variables [17]. In this way, more powerful learning models are obtained, when the FS model selects subsets of variables that jointly have good predictive capacity.

A refined division of FS methods, especially applied to the latter defined group, is commonly used. They are often divided into filters, wrappers and embedded methods. When variables are selected according to data characteristics (e.g. low variance or correlated variables) they correspond to filter-type FS methods. Wrappers utilize a learning machine technique of interest as a black box, as a pre-processing step, to score subsets of variables in terms of their predictive ability. Finally, embedded methods carry out FS in the process of the training of a learning method and are usually tailored to the applied learning method [17], [18].

A wrapper-based FS method generally consists of two parts: the objective function, which may be a learning (regression or classification) method and a searching function that selects variables to be evaluated by the objective function. The results of the learning method are used to guide the searching procedure in the selection of descriptors. Consequently, the selection procedure is closely tied to the learning algorithm used, whether in quality of selection or execution time. For instance, we may get very different behaviors whether we are using linear models or nonlinear techniques [18].

2.1 Feature Selection Applied to QSAR

Many several papers successfully applied the FS strategy in bioinformatics related areas, like: drug discovery, QSAR and gene expression patterns analysis. We decided to apply descriptor selection in our work in order to detect which and how many descriptors are the most useful ones for the prediction of logP. We agreed on the use of GAs as the searching function, given that they offer a parallel search of solutions, potentially avoiding local minima. Moreover, with a correct design of a fitness function, GA inherently guides the different generations of individuals to a good if not optimal solution. In this context, the objective function corresponds to the function used for the fitness of GA.

In this way, and as a result of the review about the related work in the area, we found some inspiring papers. In ref. [9], [18], [19], [20], [21], [22] different fitness functions are tested within a GA to determine a subset reduction. In [18], [23], [24] FS is applied using a neural network (NN) for the fitness function. However, we find that this proposal has the drawback of the great amount of time required by the NN for training and thus the execution time becomes prohibitive when the number of combination of feasible selections is large.

3 Wrapper Method

We implemented a GA for searching the space of the multiple feasible selections. We propose three appropriate fitness functions for guiding the search of GA, namely: decision trees, k-nearest neighbors (KNN) and a polynomic non linear function. According to the previous classification, our proposed FS method belongs to a wrapper method because statistical or machine learning methods are used in the fitness function for assessing the prediction capability of the selected subset.

3.1 Main Characteristics of GA

Binary strings are used to represent the individuals. Each string of length m stands for a feasible descriptor selection, where m is the number of considered descriptors. A nonzero value in the ith bit position means that the ith descriptor is selected. We have constrained to a model where p bits are active for each individual. In other words, each chromosome encodes its choice of the p selected descriptors.

The initial population is randomly generated by imposing the described restriction of exactly p active descriptors on each individual. A one-point crossover is used for the recombination [25]. Non feasible individuals could take place after crossover, because the number of nonzero bits may be different than p. This problem is solved by randomly setting or resetting bit locations as needed to be up to p active bits. Since the crossover scheme inherently incorporates bit-flip mutation, we abstained to use an additional scheme of mutation.

We did different experiments and we concluded that tournament method is appropriate for the selection of parents. Furthermore, this method is preferred than others because it is particularly easy to implement and its time complexity is $O(n)$ [25]. We also included elitism, which protects the fittest individuals in any given generation, by moving them to the next generation.

3.2 Fitness Function

Taking into account that the GA objective is to determine the most relevant set of p descriptors for predicting a physicochemical property, the fitness function should estimate the accuracy of a prediction method when only the p descriptors are used. In particular, the general form of the fitness function employed is presented in the equation 1. This formula computes the mean square error of prediction (MSE):

$$F(\mathcal{P}_{Z_{1,k}}, Z_{2,k}) = \frac{1}{n_2} \sum_{(x_i, y_i) \in Z_{2,k}} (y_i - \mathcal{P}_{Z_{1,k}}(x_i))^2 . \tag{1}$$

Where:

- Z is a matrix that represents a compound dataset, where each row and column corresponds to a compound and a descriptor respectively. The last column of Z stores the experimental target values for each compound. This column vector is denoted as y.
- $\mathcal{P}_Z$ is a statistical method trained with the dataset Z. In the same way, $\mathcal{P}_Z(x)$ is the output for the $\mathcal{P}_Z$ method when the case x is presented.
- Z_1 and Z_2 are compound databases used as learning and validation sets respectively with corresponding sizes $n_1 \times m$ and $n_2 \times m$.
- $Z_{j,k}$ is a filtered dataset in accordance with the descriptor selection encoded by the kth individual. In other words, $Z_{j,k}$ only contains those variables of Z_j whose values in the corresponding locations of the kth individual's chromosome are 1.
- x_i is a vector that represents the values of the descriptors for the ith compound of a given dataset.
- y_i is the target value for the ith compound of a given dataset.

The first argument of the fitness function is the statistical method applied to a given learning set, while the second argument corresponds to a validation set, from where fitness value is calculated. In this work, three different predictor techniques were tested. The first one corresponds to decision trees (DT) (as regression trees) using Gini's diversity index for the splitting criteria and without using any kind of pruning [26]. The second is KNN regression as used in ref [9]. Both methods are local and usually applied for prediction or for FS purposes [27].

A non linear regression model was also applied in this paper as the first argument of the fitness function. A nonlinear expression is established where their coefficients ($\beta_{i,j}$) are adjusted with a nonlinear least-squares fitting by the Gauss-Newton method [28]. The corresponding and nonlinear regression model formula is presented in Equation 2, where x_i corresponds to the value of the ith descriptor for any given compound. Non linear models are not generally applied given that they need the construction of a mathematical formula. Nevertheless, we propose it as an alternative for

NN, so that non linear regressions could be carried out. It is worth mentioning that this approach circumvent the necessity of a manual tuning of the architecture and training parameters as is the case with NN.

$$\sum_i^p \sum_{j=1}^4 \beta_{i,j} x_i{}^j + \beta_0. \tag{2}$$

4 Methodology and Analysis of Results

Our proposal consists in the search of a selection of descriptors that minimizes the prediction error when they are used as input of a predictor method. This selection is fulfilled with the GA previously described. Moreover, a fair comparison is intended to be established in order to determine which fitness function works best with GA. It is worth mentioning that, as well as minimizing error, it is important to obtain relevant descriptors in a subset of minimal size.

4.1 Data Sets

Our FS method was applied to a data set of 440 organic compounds compiled from the literature [29] where their logP values at 25°C conform the modeled target variable. The choice of the data set was supported by the possibility of comparison with the previous work and also for the heterogeneous compounds that it comprises (e.g. hydrocarbons, halogens, sulfides, anilines, alcohols, carboxylic acids amongst others).

Each compound was characterized by 73 molecular descriptors commonly used for logP [30], [31], [32]. Dragon 5.4 [33] was used for calculating descriptors of the: constitutional (41), functional groups (16), properties (2) and empiricals (3) families and we completed with 11 descriptors from [29] (Table 3). Previous to the use of the data, all descriptors were normalized, so each descriptor has a standard deviation of 1.

4.2 Genetic Algorithm Parameters

In order to assess the stability of the GA in the selection and to explore the sensitivity of the choice of p in the prediction, 45 independent runs were carried out for each choice of p, where p was set to 10, 20 and 30. This same procedure was made for the three considered fitness functions, making a total of 405 runs for the GA.

The chromosome size m is 73 according to the number of calculated descriptors. For the GA runs we used typical parameter values: population size=45; crossover probability=0.8; tournament size=3, elite members=2. A phenotypic stopping criterion is used; the GA stops when the highest fitness of the population does not improve during 15 generations or when the improvement of the average fitness of the population is less than a given tolerance value.

4.3 Prediction Method

NNs are probably one of the most widely used methods for QSAR modelling [2], [6], [34]. In order to evaluate the suitability of the selection, we used a neural network ensemble (NNE) as an independent prediction method, *i.e.* it measures the accuracy of prediction for each proposed wrapper method. The number of descriptors (d) used as input for this independent prediction method is not necessarily the same as the p genes selected by the GA. With the intention of establishing a suitable (minimal cardinality and error) subset of descriptors this value was settled to 10 different values: 11, 12, 15, 20, 25, 30, 40, 50, 60 and 73. The d descriptors used for the predictor are selected from a ranking of the most selected descriptors obtained in the 45 re-runs of the GA. Each ensemble consists of three NNs, and all of them are of type feed-forward back-propagation. The specific architecture of each NN, was established according to the number picked for d . Principal Component Analysis (PCA) is applied prior to the training of the NNE, so the descriptors that contribute less than a 0.2% of the total variance are discarded and considered as redundant.

4.4 Results

With the purpose of evaluating the performance in the prediction achieved by the aforementioned fitness functions, we trained NNEs for each presented configuration of the GA and we obtained error prediction for different choices of d (Table 1, Fig. 1).

It is worth mentioning that it is not straightforward to obtain logP related works from the bibliography that allows a reproducibility or benchmarking of the results of the work, as it is the case of ref [29]. So, to enable a direct comparison with this work, the data set was identically divided into training, validation and test set, also using the same compounds in each set.

Our results were obtained after several different NN configurations and replicas, and the tendency was rather similar. Each reported error is an average over 5 replicas (15 NNs) applied to the test set.

In comparison with the backpropagation NN proposed in ref. [29], which obtains a 0.23 MAE and where similar conditions apply, our model of NNE with the assistance of the FS method has improved the accuracy of logP prediction, even when using one less descriptor (NL, p =10 fitness function).

Decision trees as fitness function have a better behavior in their variant with $p = 20$ descriptors, but with few descriptors for d , the performance is quite far from optimal. In the case of KNN, it looks like few descriptors for p is not appropriate at least when less than 20 descriptors are used for the NNE. For NL, the behavior is quite good when $p = 10$. As expected, in all models similar results are obtained when more than 25 descriptors are considered for d .

Considering the best alternative of p for each fitness function (Fig. 1 (d)) we highlight the performance of NL. It has a roughly equal behavior along all d values and takes a minimal prediction error when $d = 25$. KNN's behavior is similar to NL, except for the lowest values of d . In the case of the DT-based predictions, although they have a better performance than the previous two cases for large d , the bad performance with small d values, makes it not so valuable as an FS technique, at least for the present example.

Table 1. Prediction errors in terms of MAE, MSE and variance on 5 runs

	Method	d = 11	d = 12	d = 15	d = 20	d = 25	d = 30	d = 40	d = 50	d = 60	d = 73
MAE	NL, p = 10	0,1972	0,1922	0,1896	0,1928	0,1708	0,1720	0,1737	0,1855	0,2021	0,1840
MAE	DT, p = 20	0,2331	0,2362	0,2428	0,2114	0,1928	0,1742	0,1656	0,1626	0,1733	0,1847
MAE	KNN, p = 30	0,2562	0,2771	0,2986	0,2325	0,2079	0,1866	0,1925	0,1782	0,1827	0,1858
MSE	NL, p = 10	0,0740	0,0658	0,0738	0,0773	0,0566	0,0551	0,0593	0,0705	0,0895	0,0650
MSE	DT, p = 20	0,1013	0,1160	0,1173	0,0903	0,0745	0,0539	0,0522	0,0510	0,0521	0,0676
MSE	KNN, p = 30	0,1166	0,1266	0,1510	0,1066	0,0781	0,0730	0,0706	0,0613	0,0664	0,0635
Variance	NL, p = 10	1,330E-04	1,425E-04	7,278E-05	3,581E-04	6,015E-05	5,289E-05	8,314E-05	6,531E-05	2,197E-05	1,154E-04
Variance	DT, p = 20	4,437E-05	1,384E-04	2,064E-04	2,821E-04	7,206E-05	4,601E-05	9,563E-05	6,871E-05	3,635E-04	1,440E-04
Variance	KNN, p = 30	2,222E-04	1,079E-04	4,144E-04	6,366E-05	6,142E-05	1,032E-04	7,277E-05	1,880E-04	9,916E-05	6,228E-05

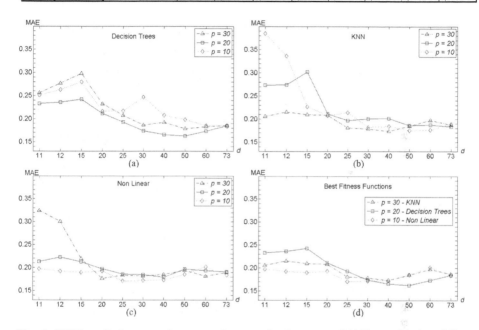

Fig. 1. NNE prediction error in terms of mean absolute error (MAE) considering different number of descriptors as input, and also for different GA-based selection methods: (a) decision trees, (b) k-nearest neighbors and (c) non linear (d) best fitness functions

Table 2. Two-way ANOVA for MAE of prediction of the three best methods and when few descriptors are used ($d = 11$, $d = 12$ and $d = 15$)

Source of Var.	Sum of Squares	D.F.	M.S.	F	p
BETWEEN	0,015629	8	0,0019536	14,7950419	2,622E-09
d factor	0,000056	2	0,0000279	0,21117041	0,8106
wrapper factor	0,015003	2	0,0075014	56,8094222	7,306E-12
Interaction	0,000570	4	0,0001426	1,0797875	0,3809
WITHIN	0,004754	36	0,0001320		
TOTAL	0,020383	44			

In order to formally support preceding facts, we analyze whether significant discrepancies exist among the different models by using a two-way ANOVA test (Table 2). The two involved factors are the FS method and the choice of d. Our comparison is focused on finding significant differences on the methods when using few descriptors

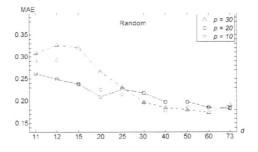

Fig. 2. NNE prediction error in terms of mean absolute error (MAE) considering different number of descriptors as input, and also for different random-based selections

Table 3. List of ranked descriptors according to wrapper method NL, $p = 10$. Descriptors with * are scaled on carbon atoms.

Rank	Symbol	Definition	Freq.	Rank	Symbol	Definition	Freq.
1	nHDon	number of donor atoms for H-bonds (N and O)	87%	38	nR06	number of 6-membered rings	42%
2	Hy	hydrophilic factor	84%	39	nR03	number of 3-membered rings	40%
3	D_S	total dipole (sum = point charge + hybridization)	78%	40	nAT	number of atoms	38%
4	nCp	number of total primary C(sp3)	78%	41	nR10	number of 10-membered rings	38%
5	D_H	total dipole (hybridization)	73%	42	nCq	number of total quaternary C(sp3)	38%
6	nH	number of Hydrogen atoms	73%	43	nCIR	number of circuits	36%
7	nSK	number of non-H atoms	71%	44	nRSR	number of sulfides	36%
8	RBN	number of rotatable bonds	71%	45	SCBO	sum of conventional bond orders (H-depleted)	33%
9	nC	number of Carbon atoms	71%	46	nCXr	number of X on ring C(sp3)	33%
10	nCs	number of total secondary C(sp3)	71%	47	nDB	number of double bonds	31%
11	nCaR	number of aromatic C(sp2)	69%	48	nBT	number of bonds	29%
12	D_P	total dipole (point charge)	67%	49	nS	number of Sulfur atoms	27%
13	IP	ionization potential	67%	50	nR09	number of 9-membered rings	24%
14	VMC1	first-order valence molecular connectivity index	67%	51	nOHs	number of secondary alcohols	24%
15	nF	number of Fluorine atoms	67%	52	nBnz	number of benzene-like rings	22%
16	Sp	sum of atomic polarizabilities [*]	64%	53	nSH	number of thiols	22%
17	Ms	mean electrotopological state	64%	54	nR=CX2	number of R=CX2	22%
18	nBO	number of non-H bonds	64%	55	nR07	number of 7-membered rings	20%
19	Ui	unsaturation index	64%	56	Mv	mean atomic van der Waals volume [*]	18%
20	nX	number of halogen atoms	62%	57	Mp	mean atomic polarizability [*]	18%
21	nHAcc	number of acceptor atoms for H-bonds (N, O, F)	62%	58	nTB	number of triple bonds	18%
22	Sv	sum of atomic van der Waals volumes [*]	60%	59	nR=CHX	number of R=CHX	18%
23	nO	number of Oxygen atoms	60%	60	nI	number of Iodine atoms	16%
24	VMC2	second-order valence molecular connectivity index	58%	61	ARR	aromatic ratio	16%
25	nCL	number of Chlorine atoms	58%	62	nR11	number of 11-membered rings	13%
26	VMC4	fourth-order valence molecular connectivity index	56%	63	PSA	fragment-based polar surface area	11%
27	nAB	number of aromatic bonds	56%	64	nOHt	number of tertiary alcohols	7%
28	nBM	number of multiple bonds	53%	65	Me	mean atomic Sanderson electroneg. [*]	2%
29	nROR	number of ethers (aliphatic)	53%	66	MW	molecular weight	0%
30	nCt	number of total tertiary C(sp3)	51%	67	E2	total two-center energy	0%
31	nOHP	number of primary alcohol	51%	68	EX	exchange energy (two-center term)	0%
32	Se	sum of atomic Sanderson electroneg. [*]	49%	69	ELC	total electrostatic interaction (two-center term)	0%
33	nCIC	number of rings	49%	70	PO	average polarizability	0%
34	nBR	number of Bromine atoms	49%	71	Ss	sum of Kier-Hall electrotopological states	0%
35	nN	number of Nitrogen atoms	47%	72	RBF	rotatable bond fraction	0%
36	AMW	average molecular weight	42%	73	MR	Ghose-Crippen molar refractivity	0%
37	nR05	number of 5-membered rings	42%				

for the NNE ($d = 11, d = 12$ and $d = 15$). Given that there is not strong evidence of an interaction factor, we can separately analyze both factors. The ANOVA test shows that there is no evidence of differences on using 11, 12 or 15 descriptors for one same wrapper method (d factor near 1), and also that significant differences are found for the choice of the method for feature selection (p-value of wrapper factor ≈ 0). Finally, we also apply Bonferroni multiple comparison procedure to determine which method differs from which. With a global level of error $\alpha = 0.03$ we found that all methods differ from each other (data not shown).

Besides, in order to evidence the advantages and the differences of the application of a FS technique, we analyzed the performance when a random selection is carried out (Fig. 2). As expected, the prediction error decreases when more descriptors are considered for the NNE. On the other hand, with large d values, error is not so bad given that all descriptors are related with the target property.

Our last analysis of results is about which descriptors were selected by GA, and their frequency of selection. Table 3 shows the list of descriptors, ranked according to $NL - p = 10$ criteria, with the percentage of the times selected in the 45 runs of the GA. From a chemical perspective, it is interesting to note that the first three top-ranked descriptors are considered as reasonably influential for logP [32].

5 Conclusions

The present work proposes a methodology to detect which descriptors are the most influential to the prediction of the molecule hydrophobicity. This detection of relevant features allows a decrease in the prediction error and also a better understanding of the structure-property relationships. The key contributions of our work are the proposal of a non linear function adjusted with least squares in the fitness function and the rigorous comparison carried out by the different combinations of the wrapper variants.

Despite the unknown of the general form of the function that governs the structure-property relationship, the fourth-order polynomial function works well for the wrapper, since it captures the nonlinearity of the model, as well as it maintains an acceptable execution time performance. Besides, the GA's behavior is quite stable given the low variance of the prediction errors and the high frequency associated with the top-ranked descriptors.

According to the authors' knowledge, we did not find previous works with a ranked list of relevant features for predicting hydrophobicity. It is worth noting that in the FS step, relevant but redundant variables can be selected. However, since PCA is applied before the training of the NNE, any redundant feature is thus discarded.

Our proposal is not restricted to logP, because this method could also be applied to any physicochemical property. It would be interesting to experiment this proposal with the aggregation of other descriptor families. In this context, we are evaluating other descriptors that express interactions between functional groups in molecules. Moreover, the GA could also be developed to directly detect the most adequate number of descriptors in a multi-objective way, instead of fixing to a specific number. At this moment, we are also planning to extend the comparison with other combinations of AI methods.

References

1. Selick, H.E., Beresford, A.P., Tarbit, M.H.: The Emerging Importance of Predictive ADME Simulation in Drug Discovery. Drug Discov 7(2), 109–116 (2002)
2. Taskinen, J., Yliruusi, J.: Prediction of Physicochemical Properties Based on Neural Network Modeling. Adv. Drug Deliver. Rev. 55(9), 1163–1183 (2003)

3. Jónsdottir, S.Ó., Jørgensen, F.S., Brunak, S.: Prediction Methods and Databases Within Chemoinformatics: Emphasis on Drugs and Drug Candidates. Bioinformatics 21, 2145–2160 (2005)
4. Tetko, I.V., Bruneau, P., Mewes, H.-W., Rohrer, D.C., Poda, G.I.: Can we estimate the accuracy of ADME-Tox predictions? Drug Discov. Today 11, 700–707 (2006)
5. Huuskonnen, J.J., Livingstone, D.J., Tetko, I.V.: Neural Network Modeling for Estimation of Partition Coefficient Based on Atom-Type Electrotopological State Indices. J. Chem. Inf. Comput. Sci. 40, 947–995 (2000)
6. Agatonovic-Kustrin, S., Beresford, R.J.: Basic Concepts of Artificial Neural Network (ANN) Modeling and its Application in Pharmaceutical Research. J. Pharmaceut. Biomed. 22(5), 717–727 (2000)
7. Tetko, I.V., Livingstone, D.J., Luik, A.I.: Neural Networks Studies. 1. Comparison of Over-fitting and Overtraining. J. Chem. Inf. Comput. Sci. 35, 826–833 (1995)
8. Topliss, J.G., Edwards, R.P.: Chance Factors in Studies of Quantitative Structure-Activity Relationships. J. Med. Chem. 22(10), 1238–1244 (1979)
9. Li, L., Weinberg, C.R., Darden, T.A., Pedersen, L.G.: Gene selection for sample classifica-tion based on gene expression data: Study of sensitivity to choice of parameters of the GA/KNN method. Bioinformatics 17(12), 1131–1142 (2002)
10. Tan, T., Fu, X., Zhang, Y., Bourgeois, A.G.: A genetic algorithm-based method for feature subset selection. Soft Comput 12(2), 111–120 (2008)
11. Zhu, Z., Ong, Y., Dash, M.: Markov blanket-embedded genetic algorithm for gene selection. Pattern Recognition 40(11), 3236–3248 (2007)
12. Forman, G.: An extensive empirical study of feature selection metrics for text classification. JMLR 3, 1289–1306 (2003)
13. Lin, K., Kang, K., Huang, Y., Zhou, C., Wang, B.: Naive bayes text categorization using improved feature selection. Journal of Computational Information Systems 3(3), 1159–1164 (2007)
14. Montañés, E., Quevedo, J.R., Combarro, E.F., Díaz, I., Ranilla, J.: A hybrid feature selection method for text categorization. International Journal of Uncertainty, Fuzziness and Knowlege-Based Systems 15(2), 133–151 (2007)
15. Kohavi, R., John, G.: Wrappers for feature selection. Artificial Intelligence 97, 273–324 (1997)
16. Blum, A., Langley, P.: Selection of relevant features and examples in machine learning. Artificial Intelligence 97, 245–271 (1997)
17. Guyon, I., Elisseeff, A.: An Introduction to Variable and Feature Selection. JMLR 3, 1157–1182 (2003)
18. Dutta, D., Guha, R., Wild, D., Chen, T.: Ensemble Feature Selection: Consistent Descriptor Subsets for Multiple QSAR Models. J. Chem. Inf. Model. 47, 989–997 (2007)
19. Liu, S., Liu, H., Yin, C., Wang, L.: VSMP: A novel variable selection and modeling method based on the prediction. J. Chem. Inf. Comp. Sci. 43(3), 964–969 (2003)
20. Wegner, J.K., Zell, A.: Prediction of aqueous solubility and partition coefficient optimized by a genetic algorithm based descriptor selection method. J. Chem. Inf. Comp. Sci. 43(3), 1077–1084 (2003)
21. Kah, M., Brown, C.D.: Prediction of the adsorption of Ionizable pesticides in soils. J. Agr. Food Chem. 55(6), 2312–2322 (2007)
22. Bayram, E., Santago, P., Harrisb, R., Xiaob, Y., Clausetc, A.J., Schmittb, J.D.: Genetic algorithms and self-organizing maps: A powerful combination for modeling complex QSAR and QSPR problems. J. of Comput.-Aided Mol. Des. 18, 483–493 (2004)

23. So, S.-S., Karplus, M.: Evolutionary Optimization in Quantitative Structure-Activity Relationship: An Application of Genetic Neural Networks. J. Med. Chem. 39, 1521–1530 (1996)
24. Fernández, M., Tundidor-Camba, A., Caballero, J.: Modeling of cyclin-dependent kinase inhibition by 1H-pyrazolo[3,4-d] pyrimidine derivatives using artificial neural network ensembles. J. Chem Inf. and Model. 45(6), 1884–1895 (2005)
25. Goldberg, D.E., Deb, K.: A comparative analysis of selection schemes used in genetic algorithms. In: Foundations of Genetic Algorithms, pp. 69–93. Morgan Kaufmann, San Mateo, CA (1991)
26. Breiman, L.: Classification and Regression Trees. Chapman & Hall, Boca Raton (1993)
27. Trevino, V., Falciani, F.: GALGO: An R package for multivariate variable selection using genetic algorithms. Bioinformatics 22(9), 1154–1156 (2006)
28. Madsen, K., Nielsen, H.B., Tingleff, O.: Methods for Non-Linear Least Squares Problems. Technical University of Denmark, 2nd edn. (April, 2004)
29. Yaffe, D., Cohen, Y., Espinosa, G., Arenas, A., Giralt, F.: Fuzzy ARTMAP and back-propagation neural networks based quantitative structure - property relationships (QSPRs) for octanol: Water partition coefficient of organic compounds. J. Chem. Inf. Comp. Sci. 42(2), 162–183 (2002)
30. Linpinski, C.A., Lombardo, F., Dominy, B.W., Freeny, P.: Experimental and Computational Approaches to Estimate Solubility and Permeability in Drug Discovery and Development Settings. Adv. Drug Deliv. Rev. 23, 3–25 (1997)
31. Duprat, A., Huynh, T., Dreyfus, G.: Towards a principled methodology for neural network design and performance evaluation in qsar; application to the prediction of logp. J. Chem. Inf. Comp. Sci. 38, 586–594 (1998)
32. Wang, R., Fu, Y., Lai, L.: A new atom-additive method for calculating partition coefficients. J. Chem. Inf. Comp. Sci. 37(3), 615–621 (1997)
33. Tetko, I.V., Gasteiger, J., Todeschini, R., Mauri, A., Livingstone, D., Ertl, P., Palyulin, V.A., Radchenko, E.V., Zefirov, N.S., Makarenko, A.S., Tanchuk, V.Y., Prokopenko, V.V.: Virtual computational chemistry laboratory - design and description. J. Comput. Aid. Mol. Des. 19, 453–463 (2005)
34. Winkler, D.A.: Neural networks in ADME and toxicity prediction. Drug. Future 29(10), 1043–1057 (2004)

On the Convergence of Protein Structure and Dynamics. Statistical Learning Studies of Pseudo Folding Pathways

Alessandro Vullo[1], Andrea Passerini[2], Paolo Frasconi[2],
Fabrizio Costa[2], and Gianluca Pollastri[1]

[1] School of Computer Science and Informatics
University College Dublin, Belfield, Dublin 4, Ireland
[2] Dipartimento di Sistemi e Informatica
Università degli Studi di Firenze, Via di S.Marta 3, 50139 Firenze, Italy

Abstract. Many algorithms that attempt to predict proteins' native structure from sequence need to generate a large set of hypotheses in order to ensure that nearly correct structures are included, leading to the problem of assessing the quality of alternative 3D conformations. This problem has been mostly approached by focusing on the final 3D conformation, with machine learning techniques playing a leading role. We argue in this paper that additional information for recognising native-like structures can be obtained by regarding the final conformation as the result of a generative process reminiscent of the folding process that generates structures in nature. We introduce a coarse representation of protein pseudo-folding based on binary trees and introduce a kernel function for assessing their similarity. Kernel-based analysis techniques empirically demonstrate a significant correlation between information contained into pseudo-folding trees and features of native folds in a large and non-redundant set of proteins.

1   Introduction

Accurate protein structure prediction is still an open and challenging problem for a vast subset of the protein universe. Experiments of blind prediction such as the CASP series [14] demonstrate that the goal is far from being achieved, especially for those proteins whose sequence does not resemble that of any protein of known structure (nearly half of the total) - the field known as *ab initio*. Difficulties in this case are well known: the choice of a reduced protein representation and the corresponding empirical potential function may allow for an efficient search of the conformational space, but generally the methods are not sensitive enough to differentiate correct native structures from conformations that are structurally close to the native state. On the other hand, techniques such as Comparative Modelling and Fold Recognition can be very successful at predicting accurate models, but success strongly depends on the quality of the alignment and the ability to reliably detect homologues. Moreover, models with severely unrealistic geometry can be produced, especially when using fully

E. Marchiori and J.H. Moore (Eds.): EvoBIO 2008, LNCS 4973, pp. 200–211, 2008.

automated prediction pipelines. As past and recent findings suggest, a practical way to obtain improvements in protein structure prediction consists of the integration of alternative techniques and sources of information. For instance, empirical elements (e.g. secondary structure predictions) are routinely used to constrain the space of allowed conformations, to correct and refine an alignment or to improve the sensitivity of remote homologue detection. Model quality assessment programs (MQAPs) are becoming increasingly important for filtering out wrong predictions [17]. A common theme between computational prediction techniques and most refinement methods is that they more or less directly depend on knowledge mined from existing protein structures and, to a smaller extent, on the available theory and principles of protein structure. In spite of the continuous increase in the amount of available structural data, progresses in protein structure prediction and model quality assessment have been slow. This may indicate that the goal of reaching reliable protein structure prediction requires new, alternative sources of information.

This paper is an attempt to investigate in this direction. We believe that novel algorithmic ideas may come from looking at the dynamics of protein folding simulations, instead of focussing solely on their final product. We assume that any plausible abstraction of the folding process may contain potentially valuable information about the final fold. Indeed, specific folding patterns are intimately related with the native structure. If deviations from these pathways occur, often this will yield incorrect (i.e. non native-like) contacts between residues that are more stable than the correct ones, resulting in structural deviations from the native fold [8]. Folding may then be viewed as the dynamical fingerprint of the resulting structure.

Modelling or understanding protein folding at the conceptual level remains beyond the scope of the present paper. Theoretical modelling of the dynamics of protein folding faces several difficulties: there is a much smaller body of experimental data than the PDB, which is typically at low resolution, and carrying out computations over long time scales requires either very large amounts of computer time or the use of highly approximate models [10]. Rather, we take the more pragmatic perspective of finding manageable representations of protein pseudo-folding simulations and evaluating their potential impact on protein structure prediction. In this study, we derive a representation called binary pseudo-folding tree (BPFT), borrowing ideas from other recent works [11,22]. A BPFT expresses a hierarchy of timestamped pairing events involving secondary structure elements (SSEs) and is computed by inspecting the execution trace of a stochastic optimisation algorithm for structure reconstruction that explores a protein conformational space driven by spatial proximity restraints. Similar algorithms are common for example in the NMR structure determination literature and can be applied to recover protein structure from contact maps [18]. We empirically investigate the existence of a relationship between information provided by BPFTs and features of native folds for a large and non-redundant set of proteins. We first introduce a kernel function for measuring similarity between BPFTs, and compare its ability to detect similarities with respect to the

TM-score [23]. We then apply the kernel to cluster sets of optimisation traces associated with alternative reconstructions from contact maps.

2   Binary Pseudo-folding Trees

Although the fine mechanisms that regulate protein folding are in principle extremely complex, hence nearly impossible to simulate and predict on current computational hardware, there is evidence that the essential elements of the process are much simpler and coarse-grained [2,15]. In nature, the folding process appears to follow "pathways", involving hierarchical assemblies and intermediate states requiring doing and undoing of structures [9]. Rather than static, and driven by properties identifiable in the final fold, the folding process appears to be dynamic and driven by interactions whose nature and relative importance change during the process itself. Multiple pathways, with different transition states also appear to be possible [21]. The combination of experimental and computational techniques has revealed other properties of the folding process [7]. For instance, it appears that in some cases interactions among key elements in the protein form a core or nucleus that essentially constrains the protein topology to its fold [19]. Also, there is much evidence that folding is hierarchical; for some proteins it involves stable intermediates, called foldons, that consist of SSEs [12]. Folding routes can then be thought of as having an underlying tree structure [11] and clusters of interacting SSEs may form the tree labels [22].

Our aim is to derive representations of protein folding simulations which have to be simple yet informative, i.e. tractable by machine learning techniques. We borrow ideas from the work of other authors [11,22], although with different premises and details. Neither we assume that the three-dimensional (3D) structure of a protein is known nor we want to identify real folding pathways for the protein under study. Rather we argue that regarding a predicted protein conformation as the result of a generative process may yield additional information about this conformation. Structures are generated by an algorithm that explores the conformational space of the protein. A labelled binary tree is built in an incremental fashion by observing notable intermediate events happening along the trajectory that is being followed. Since we are not dealing with the real process, we call the trajectory a pseudo-folding pathway and the resulting tree a binary pseudo-folding tree.

2.1   Pseudo-folding Pathways

Protein folding simulations are carried out with an algorithm that models protein structures by exploring a protein's conformational space starting from an initial (random) configuration. Usually, this kind of algorithms are guided by some form of energy encoding structural principles or a pseudo-energy (statistical potential function) or a combination of the two. In this work, we employ 3Distill [3], a machine learning based system for the prediction of alpha carbon (C_α) traces. For a given input sequence, first a set of 1D features is predicted, e.g. secondary structure and solvent accessibility. These features are then used as an input to infer the shape of 2D features like the contact map (binary or multi-class). In the last stage, protein structures are coarsely described with their backbone C_α atoms and are predicted by means of a stochastic optimisation algorithm using as pseudo-energy a function of the geometric constraints inferred from the underlying set of 1D and 2D predictions. The stochastic optimisation algorithm explores the configurational space starting from a random conformation, and refining this by global optimisation of the pseudo-potential function using local moves and a simulated annealing protocol. For more and complete details on the form of the cost function and the annealing protocol see [4].

2.2   Notation

Let $s_1 \ldots s_m$ be the sequence of m secondary structure segments of the protein, where s_i is the i-th segment in the sequence, either a α-helix or a β-strand. Let $S_1 \ldots S_T$ be the time ordered sequence of structures observed at discrete time steps during a simulation. Using this notation, S_1 is the initial configuration and S_T is the predicted model structure. We introduce a simple and synthetic representation of an execution trace based on binary trees.

A Binary Pseudo-Folding Tree (BPFT) is a rooted, unordered, leaf-labelled binary tree. Suppose we are given the pseudo-folding pathway $\mathcal{P} = S_1 \ldots S_T$. The corresponding BPFT, called $\mathcal{T}$, expresses a hierarchy of timestamped pairing events involving sets of α-helices and β-strands[1]. Each leaf node has a label that represents the type and position of a SSE (e.g. β_2 means the second strand of the sequence, α_1 the first helix and so on). An internal node $n \in \mathcal{T}$ corresponds to a pairing event occurred at time $1 < t < T$ and that involved two SSEs belonging to different clusters of interacting SSEs. Each of the two clusters is a child node of n, which in turn represents a larger set of SSEs that eventually joins another cluster in its parent. The recursive structure of $\mathcal{T}$ is inspired to other binary tree representations of folding pathways [22], but with a number of differences. In [22], a tree (the predicted folding pathway) is built by recursively applying a polynomial-time mincut algorithm to a weighted graph, this graph representing sets of interacting SSEs of the *known* experimental structure. Here, we do not assume to know the real 3D structure of a protein, unless we run the 3Distill reconstruction algorithm with experimental 1D and 2D restraints. Moreover, folding information is obtained by using a pseudo-folding trajectory, i.e. simulated dynamical data. For a given time step t of the simulation there is a node $n \in \mathcal{T}$ such that the subtree $\mathcal{T}_n$ rooted at n corresponds to the assembling history (from $t = 1 \ldots t$) of a cluster of interacting SSEs in S_t, where the segments involved are given by the leaves dominated by n. Let $\text{ch}_l[n]$ (resp. $\text{ch}_r[n]$) be the assigned left (resp. right) child of n and let $\text{LEAVES}(\cdot)$ be a function returning the set of leaf labels of a subtree of $\mathcal{T}$. The cluster of node n is formed because one or more segments in $\text{LEAVES}(\mathcal{T}_{\text{ch}_l[n]})$ interact with segments in $\text{LEAVES}(\mathcal{T}_{\text{ch}_r[n]})$,

---

[1] Random coil fragments are not usually involved in major structure stabilisation events and are not considered here.

Fig. 1. BPFT for a protein (PDB code 1WITA) as a result of the application of Alg.1 to the trajectory followed by the reconstruction algorithm described in section 2.1

thus forming a larger cluster of pairwise interacting segments in n. An example BPFT can be seen in Fig. 1. For convenience, each internal node in the figure has a numerical index. The internal node 3 represents a cluster of interactions between the segments $\alpha_1\beta_2\beta_6$ in an intermediate fold S_t ($1 < t < T$). The cluster has formed because the first helix (α_1) started to interact with the β-sheet made by the second and sixth strands (node 1). Other portions of the tree can be similarly interpreted. The simulation ends in the predicted fold which is symbolically represented by the root node; its children indicate that the final structure was predicted by joining the first strand (β_1, left child) with one or more of the segments (i.e. leaves) dominated by node 7.

2.3   BPFT Construction Algorithm

The pseudo codes of Algorithms 1 and 2 describe the procedure that we apply to build BPFTs. Parameters of GENERATEBPFT are the set of indexed SSEs of the protein and the ordered sequence of structures found along the whole simulation trajectory, from $t = 1\ldots T$. The BPFT $\mathcal{T}$ is built bottom-up, from the leaves to the root node. The structure returned by Algorithm 1 describes the assembling history of S_T as a hierarchical set of SSEs pairing events. Steps 1 to 6 initialise the partial tree with m leaf nodes, each one representing an isolated SSE not interacting with the others. This corresponds to the initial structural configuration before the configurational search starts. New nodes are then added whenever, moving from step t to $t + 1$ of the trajectory, we find that new SSEs interactions have been formed. If we add a new node (a new potentially larger cluster), its children (subclusters) are not necessarily searched among the last added nodes, because these might not longer represents clusters in S_{t+1} (i.e. at time $t + 1$, SSEs links in S_t may have been broken as well). For these reasons, $\mathcal{T}$ maintains a reference to a subset of its nodes, the 'frontier', each node pointing to a cluster of SSE interactions that are present in the fold at the current time

Algorithm 1. GENERATEBPFT($\{s_1 \ldots s_m\}, \{S_1 \ldots S_T\}$)

1: $\mathcal{T} \leftarrow \emptyset$
2: **for** $i \leftarrow 1 \ldots m$ **do**
3:     $v \leftarrow$ CREATENODE($\{s_i\}$)
4:     ADDNODE($\mathcal{T}, v, \emptyset$)
5:     $\mathcal{T}.frontier \leftarrow \mathcal{T}.frontier \cup \{v\}$
6: **end for**
7: $\mathcal{C}_T \leftarrow$ contact map of S_T
8: $\mathcal{NC} \leftarrow \emptyset$ {*Contacts of S_T formed so far*}
9: **for** $t \leftarrow 1 \ldots T$ **do**
10:     $(\mathcal{C}_t, \mathcal{CC}_t) \leftarrow$ (residue, coarse) contact maps of S_t
11:     $\mathcal{NC}_t \leftarrow \mathcal{C}_t \cap \mathcal{C}_T$ {*Contacts of S_T in current fold*}
12:     **if** $\mathcal{NC}_t \setminus \mathcal{NC} \neq \emptyset$ **then**
13:         UPDATETREE($\mathcal{T}, \mathcal{CC}_t$) {*Update tree if there are new native contacts*}
14:     **end if**
15:     $\mathcal{NC} \leftarrow \mathcal{NC}_t$ {*Update the set of temporarily formed native contacts*}
16: **end for**

---

step. Whenever we add a new node, its cluster must describe pairings between smaller subclusters of the current fold, so that the children are always searched among the frontier nodes. At time step 0, the structure is assumed to contain only isolated segments (not forming any interaction), so that the frontier is made with only leaf nodes (Step 5). In order to build and complete the tree, the trajectory is monitored searching for events that involve SSEs interactions. This is accomplished by looking, at each step, at the formation of contacts among residues in different SSEs, with the constraint that these contacts exist in the final predicted fold S_T. We motivate this choice from the assumption that the topology of the protein, here represented by the contact map $\mathcal{C}_T$ of S_T in Step 7, has an influence on the corresponding pathway [1]. In Step 8, $\mathcal{NC}$ keeps trace of the set of contacts of S_T formed until a given time step of the simulation. From Step 9 to 16, the algorithm analyses the structure S_t of each time step t of the trajectory: $\mathcal{NC}_t$ is assigned to the set of contacts of S_T formed in S_t (Step 11) and if new contacts are formed with respect to those formed in steps $1 \ldots t - 1$ (step 12), the tree is updated by a call to UPDATETREE (Step 13) passing as parameter the coarse contact map of $S_t{}^2$. Alg. 2 first updates $\mathcal{T}$'s frontier such that its nodes correctly represent clusters of SSE interactions of the last visited structure (steps 1-6). For each frontier node n, segments in LEAVES($\mathcal{T}_n$) form the vertexes of a graph with edges between interacting SSEs in the last coarse contact map. The nodes are partitioned into subsets of pairwise interacting SSEs[3] (Step 2). If there is only one component, the segments of n represent a portion of the interactions in the last fold. Hence the node is still in the frontier and will be

---

[2] A coarse contact map represents SSEs interactions and is defined similarly to a residue contact map: SSEs are used instead of residues, see e.g. [16].

[3] PARTITION($\cdot$) is implemented by computing the connected components of the graph using a simple depth first search.

Algorithm 2. UPDATETREE($\mathcal{T}$, $\mathcal{CC}$)

1: **for** $n \in \mathcal{T}.frontier$ **do**
2:    $C \leftarrow$ PARTITION(LEAVES($\mathcal{T}_n$), $\mathcal{CC}$)
3:    **if** $|C| > 1$ **then**
4:       $\mathcal{T}.frontier \leftarrow$ UPDATEFRONTIER($\mathcal{T}$, n, C)
5:    **end if**
6: **end for**
7: **for** $(s_i, s_j) \in \mathcal{CC}$ **do**
8:    $v \leftarrow \{x \in \mathcal{T}.frontier \mid s_i \in \text{leaves}(\mathcal{T}_x)\}$
9:    $w \leftarrow \{x \in \mathcal{T}.frontier \mid s_j \in \text{leaves}(\mathcal{T}_x)\}$
10:    **if** $v \not\equiv w$ **then**
11:       $n \leftarrow$ CREATENODE()
12:       ADDNODE($\mathcal{T}$, n, $\{v, w\}$) {LEAVES($\mathcal{T}_n$) = LEAVES($\mathcal{T}_v$)$\cup$ LEAVES($\mathcal{T}_w$)}
13:       $\mathcal{T}.frontier \leftarrow \mathcal{T}.frontier \cup \{n\} \setminus \{v, w\}$
14:    **end if**
15: **end for**

---

searched for the next pairing operations. If this is not the case, the frontier is updated by a call to UPDATEFRONTIER (not shown) where $\mathcal{T}_n$ is visited and n is replaced by its first descendants that contain the clusters in C. In steps 7-14, we search for SSE interactions in the current fold (given by $\mathcal{CC}$) that are not represented by the partial tree built so far. The frontier nodes are searched for those containing two interacting SSEs (steps 8-9). If the corresponding nodes are distinct, it means that no node in $\mathcal{T}$ encodes the interaction so that a new node is formed as a parent of the two nodes; the frontier is updated accordingly.

2.4   Mining Frequent Pseudo-folding Patterns

We briefly discuss an efficient procedure used to capture simple descriptions of the dominant features of pseudo-folding simulations, as represented by BPFTs, and then compare these descriptions with known experimental folding facts of a set of proteins considered in previous studies [22]. In this way, we test the protocol for its ability to mimic the real folding process.

We wish to discover patterns in pseudo-folding pathways represented by BPFTs. Since the simulator is stochastic, given the same set of restraints, any two runs could output different BPFTs varying both in shape and size. To tackle this, we represent a pseudo-folding landscape by the distribution of labelled subtrees in pseudo-folding pathways represented as BPFTs. Patterns can be naturally thought of as being the common subtrees of a set of BPFTs. We search for these patterns by mining the most frequent subtrees [5]. We have applied the methodology described to the set of proteins considered in [22]. For each protein, the reconstruction algorithm ran 200 times with the restraints defined by the native contact map, thus obtaining a sample of possible trajectories, hence BPFTs, leading to the correct native structure. From these trees we mined the most frequent subtrees and compared the events they describe with known facts about the folding of the protein under study. We have found significant

Table 1. Top 5 most frequent sub BPFTs mined from a sample of reconstruction traces (chain 1O6XA). Each subtree's support is the normalised frequency wrt to sample size.

Rank	Support	SubBPFT
1	0.85	$\beta_1\beta_2$
2	0.70	$(\alpha_2(\beta_1\beta_2))$
3	0.69	$(\beta_3(\alpha_2(\beta_1\beta_2)))$
4	0.63	$(\alpha_1(\alpha_3(\alpha_2(\beta_1\beta_2))))$
5	0.14	$(\beta_2(\beta_3(\beta_1\beta_2)))$

correspondences between our artificial samples and the experimental evidence. Most of the events described in the literature appear as encoded in one or more of the most frequent subtrees. For instance, Table 1 shows the top five frequent subtrees for one of the chains under study (PDB code 1O6XA). It is known that the folding nucleus of 1O6X is made by packing of the second helix with the β-sheet formed by $\beta_2\beta_1$ [22]. Indeed, we found the second most frequent subtree $(\alpha_2(\beta_1\beta_2))$ as perfectly describing this event, where the most frequent subtree indicates the formation of the β-sheet $\beta_2\beta_1$.

3   Kernels on BPFT

We develop kernels (i.e. similarity measures) between BPTFs to investigate the informative content of the proposed features by learning techniques. For efficiency issues, we turn BPFTs into ordered trees, by imposing a total order on the leaves according to the relative position of the SSEs in the protein sequence. We focus only on *complete* subtrees, that is subtrees that contain all descendants of the subtree root up to the leaves of the original tree. We can now apply a set kernel on complete subtrees by decomposing each BPFT into the set of its complete subtrees, and comparing two BPFTs by summing up all pairwise comparisons between elements of the two sets:

$$K(\mathcal{T},\mathcal{T}') = \sum_{n\in\mathcal{T}} \sum_{m\in\mathcal{T}'} k(\mathcal{T}_n,\mathcal{T}'_m) \tag{1}$$

To keep things simple, we compare subtrees by the delta function $k(\mathcal{T}_n,\mathcal{T}'_m) = \delta(\mathcal{T}_n,\mathcal{T}'_m)$. The overall kernel computes the similarity between two BPFTs by counting the number of complete subtrees (i.e. partial pseudo-folding representations) they have in common. In the following, we refer to this kernel as *cluster-node* kernel. Note that by imposing a canonical ordering to BPFTs and having no timestamps in the internal nodes, we only care of the hierarchy of interactions between SSE clusters, ignoring differences due to the relative timestamp of events involving non-overlapping clusters. Such invariance aims at modelling cases in which separate portions of a chain fold independently, a situation which is known to take place in nature. Comparison of complete subtrees of size one (i.e. leaves) provides an informative contribution whenever two simulations rely on different SSE predictions. Note that the cluster-node kernel does not retain information of temporary interactions which form during the process but are not preserved in the final structure. Moreover, the kernel compares SSE clusters, but it does not consider the specific SSE pairs responsible for the formation of a cluster, apart from those formed by exactly two SSEs.

By this, we also consider a variant where the description of internal BFPT nodes is enriched with three different sets of SSE pairs: those which began interacting when the cluster formed; those which preserved their interaction; those whose interaction was lost when the cluster formed. A new subtree kernel accounting for such information is defined as follows:

$$k(\mathcal{T}_n, \mathcal{T}'_m) = \delta(\mathcal{T}_n, \mathcal{T}'_m) + \sum_{\substack{i \in F(n) \\ j \in F(m)}} \delta(i,j) + \sum_{\substack{i \in P(n) \\ j \in P(m)}} \delta(i,j) + \sum_{\substack{i \in L(n) \\ j \in L(m)}} \delta(i,j) \quad (2)$$

where $F(n), P(n), L(n)$ represent the sets of pairwise SSE interactions which are respectively formed, preserved and lost in the cluster corresponding to node n. This kernel is dubbed *pairwise-interaction* kernel in the following. Note that the kernels described in this section are conceived for measuring similarities between BPFTs originating from simulations on the same protein sequence, even if with possibly different restraints. The extension to inter-protein similarities is subject of ongoing investigation.

4   Experiments and Discussion

Given a predicted structure and its pseudo-folding pathway, we first test whether the corresponding BFPT retains some information about the distance between the predicted and native (unknown) fold. We thus generated a data set of pseudo-folding simulations for 250 non-redundant PDB chains (maximum 25% mutual sequence similarity for any two chains) considered in [4] by running 3Distill (see Sec. 2.1) using restraints obtained from four increasingly noisy contact maps: the native one, contact maps obtained from PDB templates with a max sequence identity threshold at 95% and 50% respectively, and an *ab initio* predicted map. For each of these maps, 200 simulations were run, resulting in 800 structures for each protein. The TM-score function [23] was used to measure the distance between the predicted and native fold. BPFTs were generated from the pseudo-folding processes using Alg. 1, and the two kernels defined in Section 3 were employed to measure pairwise BPFT similarities. The kernels were normalised as suggested in [6], i.e. the input vectors are normalised in feature space and centered by shifting the origin to their center of gravity. Figures 2(a) and 2(b) show the kernel matrices obtained averaging over structures with similar quality, for the cluster-node and pairwise-interaction kernel respectively. Each $([i, i+1], [j, j+1])$ bin in the maps represents the average kernel value between two structures whose TM-score to the native is in the $[i, i+1]$ and $[j, j+1]$ interval respectively. The kernel values increase with the TM-score to the native in both cases. Interestingly, the kernels discriminate pseudo-folding simulations

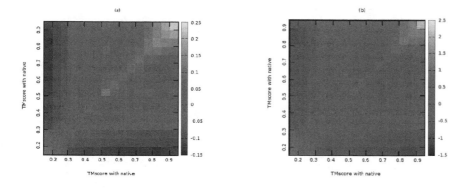

Fig. 2. Kernel matrix obtained averaging over structures with similar quality measured as TM-score with the native: (a) cluster-node kernel (b) pairwise-interaction kernel

when TM-score $\in [0.3, 0.4]$, a range of thresholds that separates poorly predicted and native-like folds [23]. This depends on the distribution of the scores, which presents a separation of the instances on the previous interval (data not shown). The kernels are clearly modelling some aspects of the given distribution.

In a binary classification setting, the relatedness of a certain kernel function to the target can be measured by the Kernel Target Alignment (KTA) [6], defined as the normalised Frobenius product between the kernel matrix and the matrix representing pairwise target products. In our setting, a binary target can be obtained using a threshold on the TM-score with respect to the native structure (we chose 0.4, see above). Figure 3 (left) reports an histogram of KTA values for our two kernels. About half of the proteins show an alignment greater than 0.15. As expected, the more informed pairwise-interaction kernel has an overall better alignment.

As a final test for the discriminative power of our two kernels, we clustered protein structures and their simulations using spectral techniques [20]. Given a matrix S of pairwise similarities between examples, they compute the principal eigenvectors of a Laplacian matrix derived from S, and apply a simple clustering algorithm, like k-means or recursive bi-partitioning, on the rows of the eigenvector matrix. As suggested in [20], we employed the multicut algorithm [13], combined with a k-means with 5 runs initialized with orthogonal centers and 20 runs initialized with random centers. Since we are mainly focussing on separation between decoys and native-like structures, the number of searched clusters was set to two. We then measured the quality of clustering using the correlation between (1) a binary value that indicates the cluster assigned to the BPFT (2) the TM-score to the native of the corresponding predicted structure. Figure 3 (right) shows histograms of the correlations obtained by clustering with the two kernels. Albeit simple, the cluster-node kernel shows a significant correlation for a large fraction of tested proteins. For 80% of the proteins, the correlation is greater than 0.15. The average correlation per protein is 0.4, and goes up to 0.47 using the more informed pairwise-interaction kernel. With this kernel we see a consistent increase of the number of cases where the correlation is more than

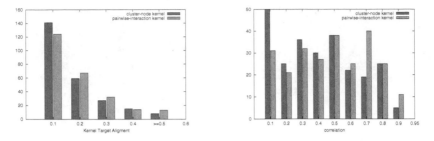

Fig. 3. Histogram of: (left) KTA values, binary targets obtained with TM-Score threshold with the native set to 0.4; (right) correlation between cluster assignment and TM-score with native structure. Results are for cluster-node and pairwise interaction kernel.

0.5. Noticeably, the ability of clustering the predicted models increases by using additional dynamical information, i.e. pairwise intermediate SSE interactions. Finally, the correlation between clustering quality and KTA value is about 0.6 for both cluster-node and pairwise-interaction kernel, thus showing a certain degree of match between the two analyses. An in-depth look at the results showed that high correlation is obtained when structures generated using the same restraints are assigned (with possibly few exceptions) to the same cluster. For the simple kernel, 42 proteins have correlation higher than 0.7. In 23 of these cases, structures generated from the native contact map are separated from all other structures, in 17 cases structures from native and 95% identity template maps are clustered together. In 1 case *ab initio* generated structures are clustered together with those from native maps, and all template-based structures are assigned to the other cluster. The last case (chain A of PDB entry 1OJH), is an interesting exception as indeed *ab initio* generated structures had a better TM-score with the native than all template-based ones.

5   Conclusions and Future Work

This study was motivated by the idea that reasonable computational abstractions of the protein folding process may contain useful information about the final protein structures. We focused on a specific pseudo-folding algorithm based on stochastic reconstruction from contact maps and empirically found that the information extracted from the pseudo-folding process does indeed allow us to define a discriminant measure of similarity (expressed by a kernel function) between the corresponding final protein structures. In particular, we found that (1) the folding abstraction used here agrees with availalble experimental evidence about the folding of some proteins, and that (2) our kernels are able to separate good and poor reconstructions of the same protein.

These findings pave the way towards the use of pseudo-folding features in the analysis and discrimination of protein structures. Attaining such a goal from a machine learning perspective requires a generalisation of the current kernel to compare pseudo-folding trees associated with different proteins.

References

1. Alm, E., Baker, D.: Prediction of protein-folding mechanisms from free-energy landscapes derived from native structures. PNAS 96, 11305–11310 (1999)
2. Baker, D.: A surprising simplicity to protein folding. Nature 405, 39–42 (2000)
3. Bau, D., Martin, A.J.M., Mooney, C., Vullo, A., Walsh, I., Pollastri, G.: Distill: A suite of web servers for the prediction of one-, two- and three-dimensional structural features of proteins. BMC Bioinformatics 7(402) (2006)
4. Bau, D., Pollastri, P., Vullo, A.: Distill: a machine learning approach to ab initio protein structure prediction. In: Bandyopadhyay, S., Maulik, U., Wang, J. (eds.) Analysis of Biological Data: A Soft Computing Approach, World Scientific, Singapore (2007)
5. Chi, Y., Nijssen, S., Muntz, R.R., Kok, J.N.: Frequent Subtree Mining–An Overview. Fundamenta Informaticæ 66(1-2), 161–198 (2005)
6. Cristianini, N., Kandola, J., Elisseef, A., Shawe-Taylor, J.: On kernel-target alignment, innovations in Machine Learning, pp. 205–256 (2006)
7. Dinner, A.R., Sali, A., Smith, L.J., Dobson, C.M., Karplus, M.: Understanding protein folding via free-energy surfaces from theory to experiments. Trends Biochem 25(7), 331–339 (2000)
8. Dobson, C.M.: The structural basis of protein folding and its links with human disease. Phil. Trans. R. Soc. Lond. 356, 133–145 (2001)
9. Dobson, C.M.: Protein folding and misfolding. Nature 426, 884–890 (2003)
10. Friesner, R.A., Prigogine, I., Rice, A.S.: Computational methods for protein folding. In: Advances in Chemical Physics, vol. 120, John Wiley, Chichester (2002)
11. Hockenmaier, J., Joshi, A.K., Dill, K.A.: Routes are trees: The parsing perspective on protein folding. Proteins 66, 1–15 (2007)
12. Maity, H., Maity, M., Krishna, M., Mayne, L., Englander, S.W.: Protein folding: the stepwise assembly of foldon units. PNAS 102, 4741–4746 (2005)
13. Meila, M., Shi, J.: A random walks view of spectral segmentation. AISTATS (2001)
14. Abstracts of the CASP7 conference, Asilomar, CA, USA, 26-30/11/ (2007), http://www.predictioncenter.org/casp7/Casp7.html
15. Plaxco, K.W., Simons, K.T., Ruczinski, I., Baker, D.L.: Topology, stability, sequence and length. Defining the determinants of two-state protein folding kinetics. Biochemistry 39, 11177–11183 (2000)
16. Pollastri, G., Vullo, A., Frasconi, P., Baldi, P.: Modular DAG-RNN architectures for assembling coarse protein structures. J. Comp. Biol. 13(3), 631–650 (2006)
17. Tosatto, S.C.: The victor/FRST function for model quality estimation. J. Comp. Biol. 12(10), 1316–1327 (2005)
18. Vendruscolo, M., Kussell, E., Domany, E.: Recovery of protein structure from contact maps. Folding and Design 2, 295–306 (1997)
19. Vendruscolo, M., Paci, E., Dobson, C., Karplus, M.: 3 key residues form a critical contact network in a protein folding transition state. Nature 409, 641–645 (2001)
20. Verma, D., Meila, M.: A comparison of spectral clustering algorithms. TR 03-05-01, University of Washington (2001)
21. Wright, C.F., Lindorff-Larsen, K., Randles, L.G., Clarke, J.: Parallel protein-unfolding pathways revealed and mapped. Nature Struct Biol 10, 658–662 (2003)
22. Zaki, M.J., Nadimpally, V., Bardhan, D., Bystroff, C.: Predicting protein folding pathways. Bioinformatics 20, i386–393 (2004)
23. Zhang, Y., Skolnick, J.: Scoring function for automated assessment of protein structure template quality. Proteins 57, 702–710 (2004)

Author Index

Lecture Notes in Computer Science

Sublibrary 1: Theoretical Computer Science and General Issues

For information about Vols. 1– 4624
please contact your bookseller or Springer

Vol. 4759: J. Labarta, K. Joe, T. Sato (Eds.), High-Performance Computing. XV, 524 pages. 2008.

Vol. 4746: A. Bondavalli, F. Brasileiro, S. Rajsbaum (Eds.), Dependable Computing. XV, 239 pages. 2007.

Vol. 4743: P. Thulasiraman, X. He, T.L. Xu, M.K. Denko, R.K. Thulasiram, L.T. Yang (Eds.), Frontiers of High Performance Computing and Networking ISPA 2007 Workshops. XXIX, 536 pages. 2007.

Vol. 4742: I. Stojmenovic, R.K. Thulasiram, L.T. Yang, W. Jia, M. Guo, R.F. de Mello (Eds.), Parallel and Distributed Processing and Applications. XX, 995 pages. 2007.

Vol. 4739: R. Moreno Díaz, F. Pichler, A. Quesada Arencibia (Eds.), Computer Aided Systems Theory – EURO-CAST 2007. XIX, 1233 pages. 2007.

Vol. 4736: S. Winter, M. Duckham, L. Kulik, B. Kuipers (Eds.), Spatial Information Theory. XV, 455 pages. 2007.

Vol. 4732: K. Schneider, J. Brandt (Eds.), Theorem Proving in Higher Order Logics. IX, 401 pages. 2007.

Vol. 4731: A. Pelc (Ed.), Distributed Computing. XVI, 510 pages. 2007.

Vol. 4728: S. Bozapalidis, G. Rahonis (Eds.), Algebraic Informatics. VIII, 291 pages. 2007.

Vol. 4726: N. Ziviani, R. Baeza-Yates (Eds.), String Processing and Information Retrieval. XII, 311 pages. 2007.

Vol. 4719: R. Backhouse, J. Gibbons, R. Hinze, J. Jeuring (Eds.), Datatype-Generic Programming. XI, 369 pages. 2007.

Vol. 4711: C.B. Jones, Z. Liu, J. Woodcock (Eds.), Theoretical Aspects of Computing – ICTAC 2007. XI, 483 pages. 2007.

Vol. 4710: C.W. George, Z. Liu, J. Woodcock (Eds.), Domain Modeling and the Duration Calculus. XI, 237 pages. 2007.

Vol. 4708: L. Kučera, A. Kučera (Eds.), Mathematical Foundations of Computer Science 2007. XVIII, 764 pages. 2007.

Vol. 4707: O. Gervasi, M.L. Gavrilova (Eds.), Computational Science and Its Applications – ICCSA 2007, Part III. XXIV, 1205 pages. 2007.

Vol. 4706: O. Gervasi, M.L. Gavrilova (Eds.), Computational Science and Its Applications – ICCSA 2007, Part II. XXIII, 1129 pages. 2007.

Vol. 4705: O. Gervasi, M.L. Gavrilova (Eds.), Computational Science and Its Applications – ICCSA 2007, Part I. XLIV, 1169 pages. 2007.

Vol. 4703: L. Caires, V.T. Vasconcelos (Eds.), CONCUR 2007 – Concurrency Theory. XIII, 507 pages. 2007.

Vol. 4700: C.B. Jones, Z. Liu, J. Woodcock (Eds.), Formal Methods and Hybrid Real-Time Systems. XVI, 539 pages. 2007.

Vol. 4699: B. Kågström, E. Elmroth, J. Dongarra, J. Waśniewski (Eds.), Applied Parallel Computing. XXIX, 1192 pages. 2007.

Vol. 4698: L. Arge, M. Hoffmann, E. Welzl (Eds.), Algorithms – ESA 2007. XV, 769 pages. 2007.

Vol. 4697: L. Choi, Y. Paek, S. Cho (Eds.), Advances in Computer Systems Architecture. XIII, 400 pages. 2007.

Vol. 4688: K. Li, M. Fei, G.W. Irwin, S. Ma (Eds.), Bio-Inspired Computational Intelligence and Applications. XIX, 805 pages. 2007.

Vol. 4684: L. Kang, Y. Liu, S. Zeng (Eds.), Evolvable Systems: From Biology to Hardware. XIV, 446 pages. 2007.

Vol. 4683: L. Kang, Y. Liu, S. Zeng (Eds.), Advances in Computation and Intelligence. XVII, 663 pages. 2007.

Vol. 4681: D.-S. Huang, L. Heutte, M. Loog (Eds.), Advanced Intelligent Computing Theories and Applications. XXVI, 1379 pages. 2007.

Vol. 4672: K. Li, C. Jesshope, H. Jin, J.-L. Gaudiot (Eds.), Network and Parallel Computing. XVIII, 558 pages. 2007.

Vol. 4671: V.E. Malyshkin (Ed.), Parallel Computing Technologies. XIV, 635 pages. 2007.

Vol. 4669: J.M. de Sá, L.A. Alexandre, W. Duch, D.P. Mandic (Eds.), Artificial Neural Networks – ICANN 2007, Part II. XXXI, 990 pages. 2007.

Vol. 4668: J.M. de Sá, L.A. Alexandre, W. Duch, D.P. Mandic (Eds.), Artificial Neural Networks – ICANN 2007, Part I. XXXI, 978 pages. 2007.

Vol. 4666: M.E. Davies, C.J. James, S.A. Abdallah, M.D. Plumbley (Eds.), Independent Component Analysis and Signal Separation. XIX, 847 pages. 2007.

Vol. 4665: J. Hromkovič, R. Královič, M. Nunkesser, P. Widmayer (Eds.), Stochastic Algorithms: Foundations and Applications. X, 167 pages. 2007.

Vol. 4664: J. Durand-Lose, M. Margenstern (Eds.), Machines, Computations, and Universality. X, 325 pages. 2007.

Vol. 4661: U. Montanari, D. Sannella, R. Bruni (Eds.), Trustworthy Global Computing. X, 339 pages. 2007.

Vol. 4649: V. Diekert, M.V. Volkov, A. Voronkov (Eds.), Computer Science – Theory and Applications. XIII, 420 pages. 2007.

Vol. 4647: R. Martin, M.A. Sabin, J.R. Winkler (Eds.), Mathematics of Surfaces XII. IX, 509 pages. 2007.

Vol. 4646: J. Duparc, T.A. Henzinger (Eds.), Computer Science Logic. XIV, 600 pages. 2007.

Vol. 4644: N. Azémard, L. Svensson (Eds.), Integrated Circuit and System Design. XIV, 583 pages. 2007.

Vol. 4641: A.-M. Kermarrec, L. Bougé, T. Priol (Eds.), Euro-Par 2007 Parallel Processing. XXVII, 974 pages. 2007.

Vol. 4639: E. Csuhaj-Varjú, Z. Ésik (Eds.), Fundamentals of Computation Theory. XIV, 508 pages. 2007.

Vol. 4638: T. Stützle, M. Birattari, H. H. Hoos (Eds.), Engineering Stochastic Local Search Algorithms. X, 223 pages. 2007.

Vol. 4630: H.J. van den Herik, P. Ciancarini, H.H.L.M.(J.) Donkers (Eds.), Computers and Games. XII, 283 pages. 2007.

Vol. 4628: L.N. de Castro, F.J. Von Zuben, H. Knidel (Eds.), Artificial Immune Systems. XII, 438 pages. 2007.

Vol. 4627: M. Charikar, K. Jansen, O. Reingold, J.D.P. Rolim (Eds.), Approximation, Randomization, and Combinatorial Optimization. XII, 626 pages. 2007.